THE STARRETT BOOK

FOR

MACHINISTS' APPRENTICES

THE STARRETT BOOK

FOR

MACHINISTS' APPRENTICES

BY

HOWARD P. FAIRFIELD
Assistant Professor Machine Construction, Worcester Polytechnic Institute

AND

CARL S. DOW, S. B.
Editor-in-chief Practical Mechanical Engineering
Editor-in-chief Practical Shop Work

ISBN: 978-93-5128-759-9(HB)

First Published in 1917
Indian Reprint in 2017

Published by

Kalpaz Publications
C-30, Satyawati Nagar,
Delhi – 110052
E-mail: kalpaz@hotmail.com
Website: kalpazpublications.com
Ph.: +91-9212142040

Cataloging in Publication Data—DK
Courtesy: D.K. Agencies (P) Ltd. <docinfo@dkagencies.com>

Fairfield, Howard P. (Parker), 1858- **author.**
The Starrett book for machinists' apprentices / by Howard P. Fairfield and Carl S. Dow (S.B.).
pages cm
Reprint. Originally published: Athol, Massachusetts : The L.S. Starrett Company, 1931.
ISBN 9789351287599 (HB)

1. Machine-shop practice—Handbooks, manuals, etc. 2. Machine-tools—Handbooks, manuals, etc. I. Dow, Carl S. (Carl Stephen), author. II. L.S. Starrett Company. III. Title.

TJ1165.F35 2017 DDC 621.8 23

INTRODUCTION

Laying out work preliminary to machining is transferring blue-print instructions on to the metal. While the blue-print gives dimensions accurately, without any great precision in the drawing itself, lines laid out on the metal are to be worked to and must therefore be accurate. No one can consider himself a skilled machinist unless he can lay out his own work and, when called upon, lay out work for the less experienced.

To become skilled in laying out should be the aim of every apprentice. Possessing this skill gives more opportunity to show ability than the running of a machine. It is a qualification one must have for advanced positions such as toolmaker, foreman, or superintendent.

But laying out requires some knowledge of mathematics, some skill at mechanical drawing, and an acquaintance with machinists' fine tools and shop operations. Attention to details and extreme care are of utmost importance. Increased labor cost, as well as material wasted because of errors in laying out, are the penalties of mistakes.

The apprentice, then, should lose no opportunity to make himself capable of laying out work. Close observation of pieces laid out by skilled machinists is one way of becoming acquainted with the art. The fortunate apprentice may also have opportunity to observe a skilled machinist while laying out various jobs.

The number of measuring and laying out tools or instruments now purchasable is very great and the apprentice must become familiar with practically all of them. He must know what he can accomplish with each so that he will instinctively select those best suited to the job in hand.

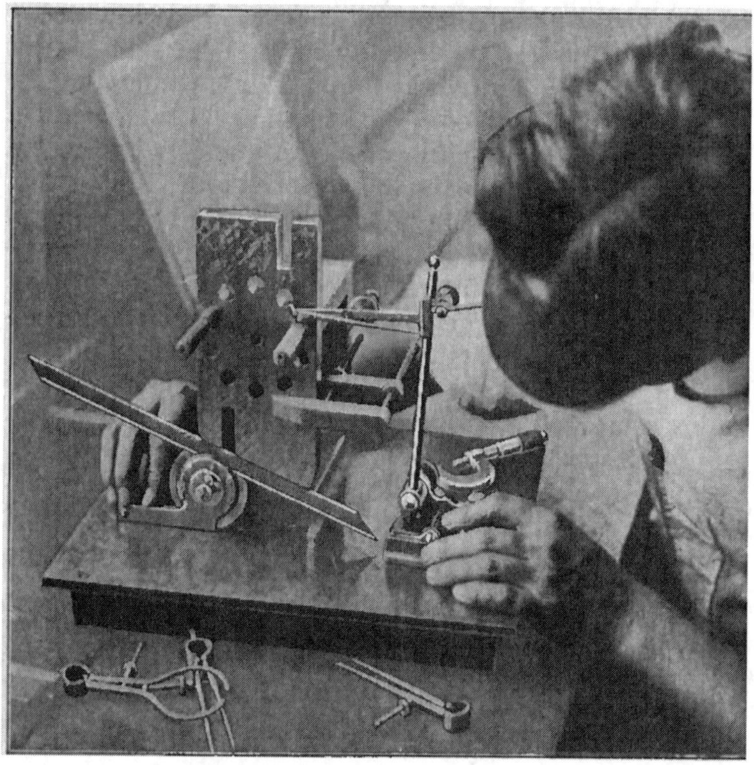

Economy of time in laying out is another element of success. Time-saving tools, such as the dial test indicator, quick-acting micrometer, and combination set, should be among those ready for use. The combination set, for instance, combines a rule, square, miter, protractor, center square, depth gage, height gage, and level. The fewer the tools used, provided the ones at hand are really good ones, the less the bench will be littered with tools which may be used only occasionally.

The tools in a machinist's tool-box are a sure indication of his ability. A well-fitted kit of fine tools helps him hold a job in hard times and is one of the best assets a man can have when applying for a job. The possession of many fine tools indicates a love for accurate work, freedom from the borrowing habit, and a *determination* to do work which will demand *recognition.* Next to having a complete outfit of fine tools is the disposition on the part of the apprentice to add the best tools as rapidly as he can afford them.

In preparing this book, the aim has been to select those elementary features most essential to the advancement of machinists' apprentices and students in technical and manual training schools. It is intended to give such students a portion of the instruction ordinarily given by the teacher or by more experienced machinists. It will also serve as a reference book for data not to be memorized.

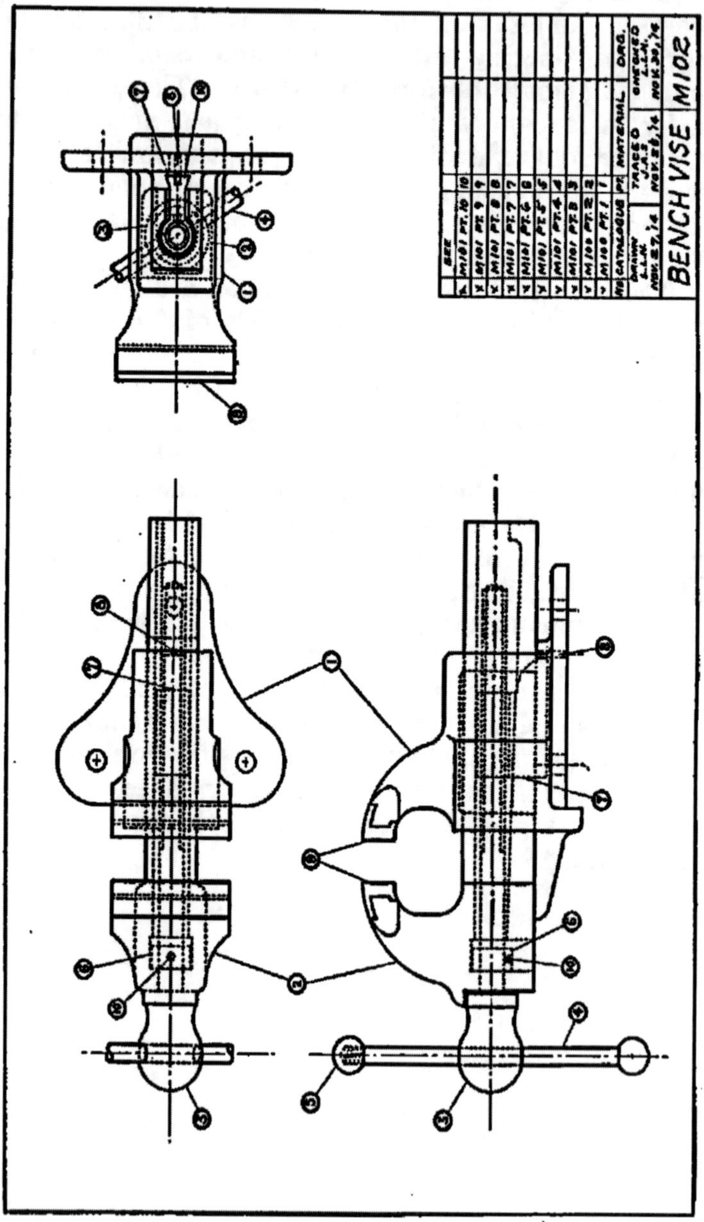

ASSEMBLY DRAWING

READING WORKING DRAWINGS

Drawing is the language of the engineer, designer, and machinist. Unless a machinist can at least read working drawings he cannot be known as a skilled mechanic. Certain conventions relating to views, lines, scales, sections, and other representations, are what make up the language of drawings, and the correct use of these is readily learned. A set of working drawings consists of

GENERAL DRAWING, showing the entire machine with all the parts located in their proper relation to one another. This drawing is usually made to a reduced scale; for example, one-quarter or one-half size; it is often termed the Assembled or Assembly Drawing.

DETAIL DRAWINGS show each part of the machine separately; they are often termed "detail," or "details." A detail drawing should be supplied with complete data for constructing the part, such as dimensions, material used, number of pieces, operations to be performed, etc., and should consist of sufficient views to be easily read. In practice some firms group several details upon a single sheet — others place a single detail upon a sheet.

SECTIONAL DRAWINGS show certain assembled portions, as if a part of the stock had been sliced away to more clearly illustrate the interior construction, often termed "sections." Position of "section" is shown by a full line drawn through a "view" and lettered at each end.

BOLT AND SCREW LISTS. On these are tabulated all bolts, screws, etc., which are common to the stockroom, and necessary to the erecting of the machine.

MOTION DIAGRAMS. Instruction is sometimes necessary concerning the relation of certain centers to the motion of parts, velocity ratios, and direction of motion; therefore where a machine has a number of more or less complicated motions, motion diagrams are provided.

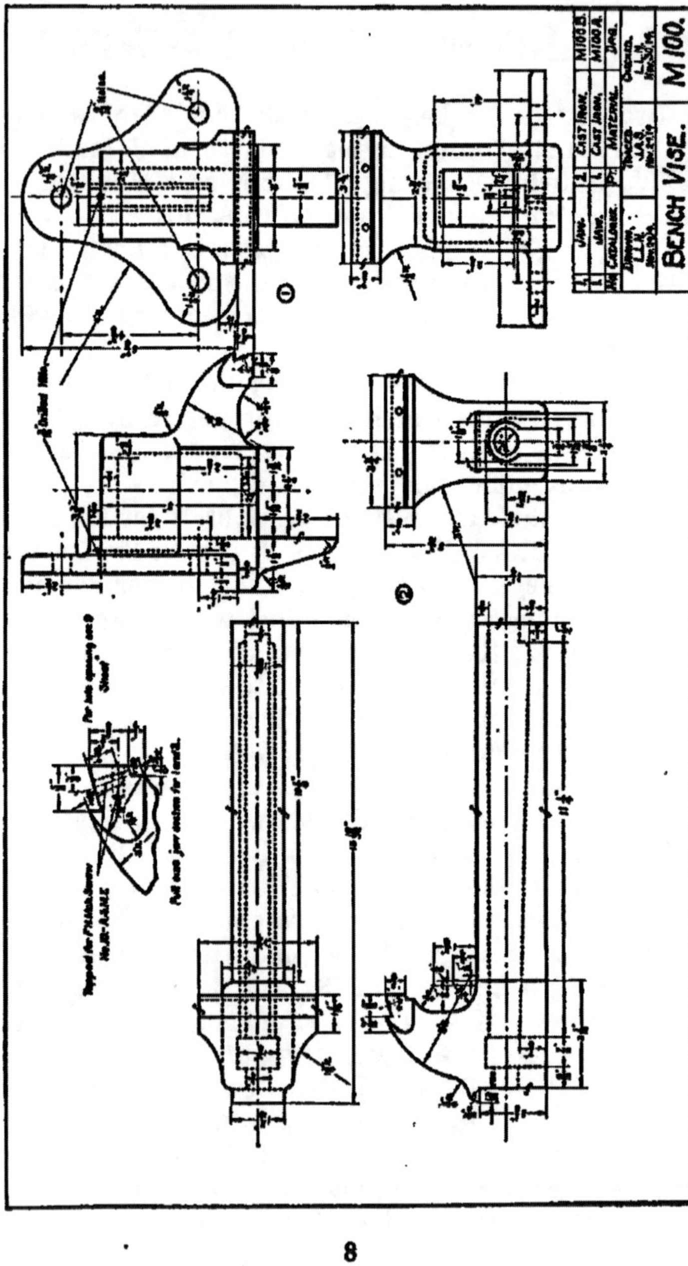

WORKING DRAWING—DETAILS

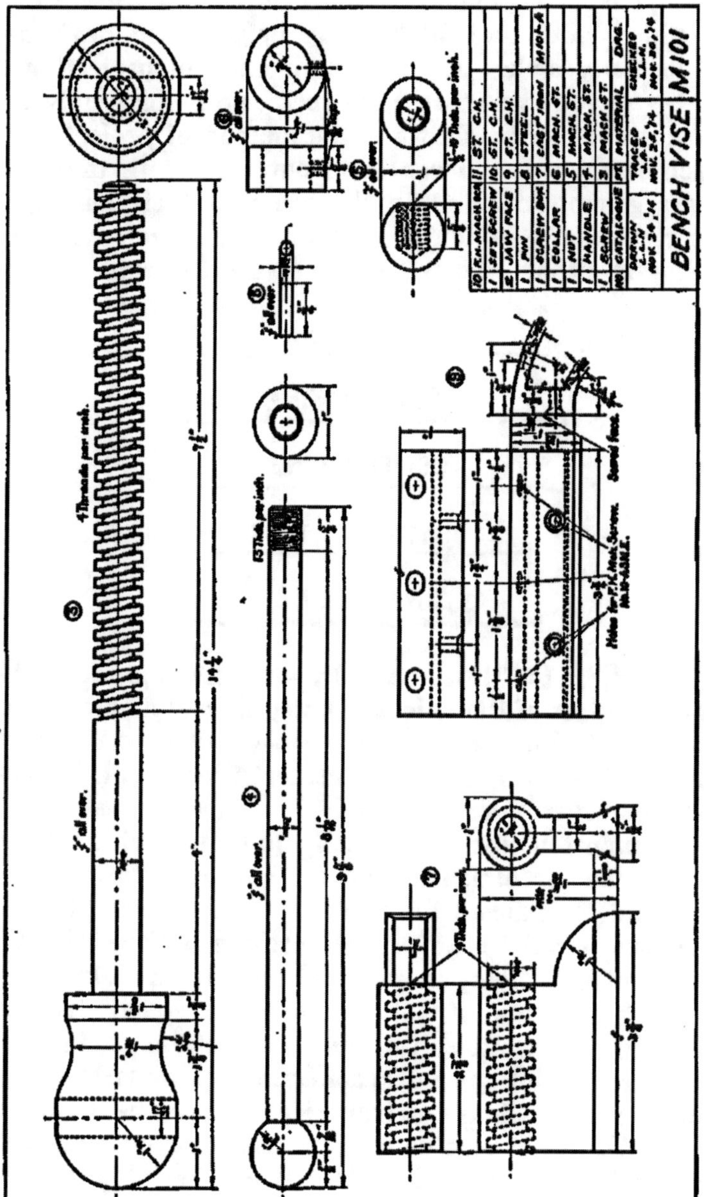

WORKING DRAWING—DETAILS

VIEWS. All material things have three dimensions; length, breadth, and thickness or height. The draftsman of necessity makes use of some method of projection to get his various views on a flat surface on which only two dimensions can be shown — the method of projection in machine-shop use places the front view with the other views grouped around in the order of their names, as top view above, bottom view below, etc.; each view centering on either a horizontal or a vertical center line.

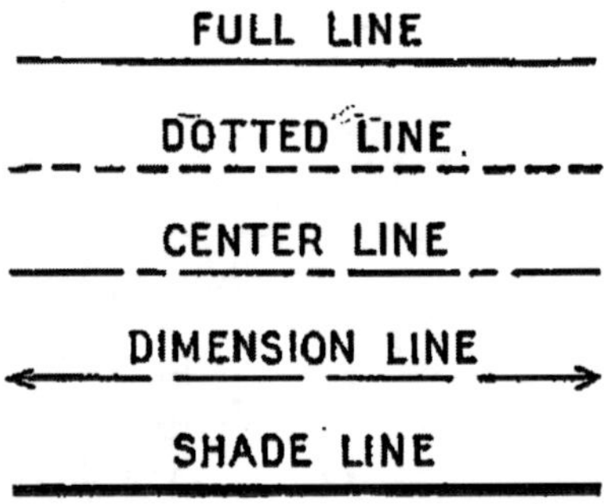

LINES. Full lines on a drawing indicate the visible lines or edges of the object. Dotted lines indicate hidden or invisible lines and edges. Broken lines, made up of dots and dashes, indicate center lines. All lay-outs should start from center lines.

Dimension lines are usually full lines with a break in the line for dimension figures and an arrow head at each end to indicate the surfaces dimensioned. Section lines are parallel lines drawn across a surface which is represented as being in section; they are usually drawn at an inclination of 45° or 60°, and equally spaced. By using for sections various combinations of full and dotted lines and special spacings, different materials of construction, such as cast iron, steel, etc., can be indicated.

SCALES. Where convenient, all drawings are made actual size, termed full scale. When the object is too large to be conveniently represented full size, the drawing is made to a regularly reduced size, called a reduced scale drawing. The usual scales are full-size, half-size, quarter-size, and eighth-size, also known as 12", 6", 3", and 1½" to 1 foot. When working from drawings the dimension figures should be invariably followed — measurements should not be taken from the drawing.

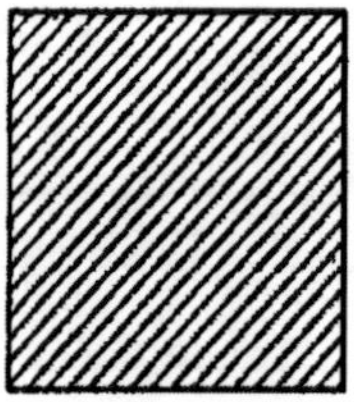

CAST IRON

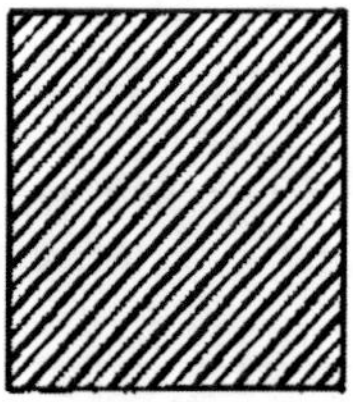

STEEL

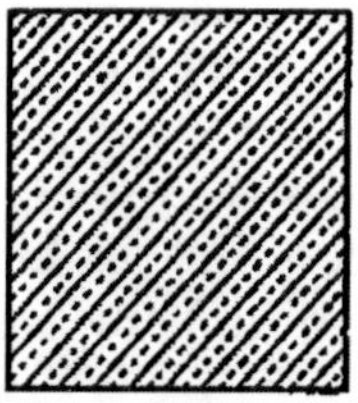

COPPER

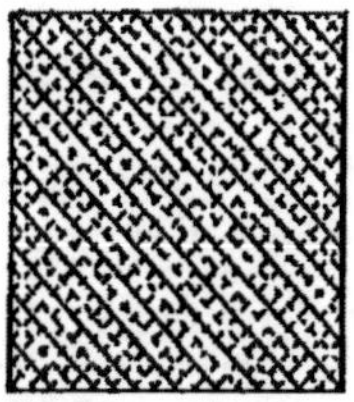

BRASS OR BRONZE

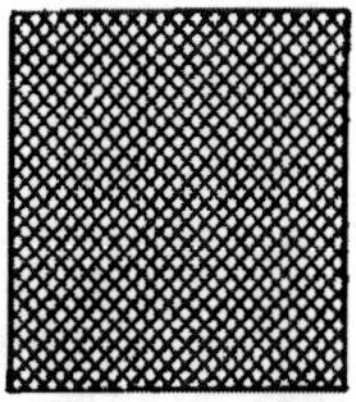

WHITE ALLOYS

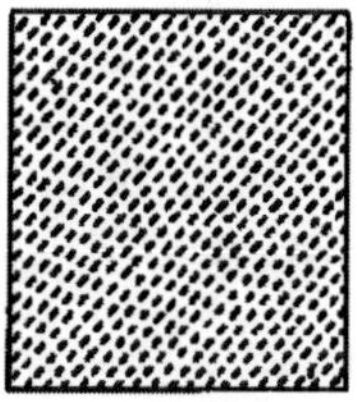

ALUMINUM

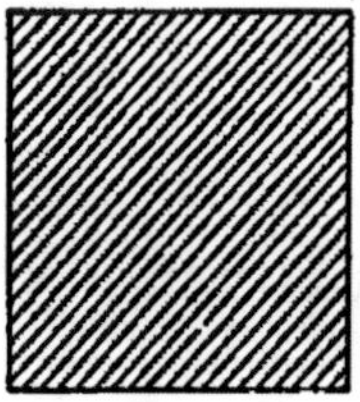

LEAD

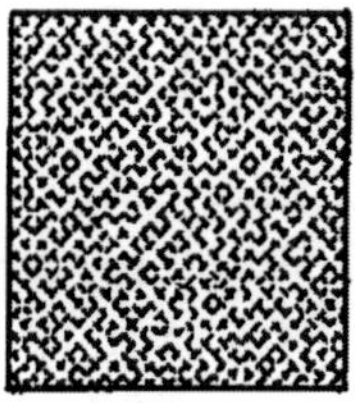

ZINC

ABBREVIATIONS. All information on a drawing is, when possible, abbreviated as follows:

CONVENTIONAL ABBREVIATIONS

Finish: Surface is to be finished	Scrape: Surface is to be hand-scraped	R. H.: Right Hand
Grind: Surface is to be ground	': Feet	L. H.: Left Hand
Face: To square up	": Inches	W. I.: Wrought Iron
Bore: Use of boring tools or bars	Dia.: Diameter	C. I.: Cast Iron
Ream: Hole should be reamed	Rad.: Radius	M. S.: Machine Steel T. S.: Tool Steel C. R. S.: Cold Rolled Steel
Tap: Hole is to be tapped	Thd.: Thread	C. S.: Carbon Steel H. S. S.: High Speed Steel.
Drill: Hole is to be drilled	U. S. S.: United States Standard	Running Fit, Drive Fit, Force Fit, Shrink Fit, Taper Fit: Allowances to be made in size of shaft

SCREW THREADS, STRUCTURAL RIVETING, PIPE FITTINGS, LINE SHAFT BEARINGS, etc., are so standardized that conventional representations are always used by the draftsmen.

MEASURING TOOLS

Measurements in general are those of length, area, and volume. In machine-shop practice the measurement of length is the common one. This is of such importance, and many of the measurements are of such exactness, that a multitude of measuring tools are being marketed, nearly all of which are for the main purpose of obtaining linear measurements.

THE YARD. In the United States the Standard of length is the British yard, of which two copies are owned by the United States Government.

THE METER, which is the French standard of length, is also coming into use in the United States,—notably in instrument work. The meter equals 39.37 inches.

The use of measuring tools in machine work is largely confined to the thirty-sixth subdivision of the yard, or the inch. The inch is subdivided into various lengths, of which the ten-thousandth part is the shortest practical shop measurement. Measurements shorter than this are, however, common enough in scientific laboratory work.

The practical machinist and toolmaker divides his work into two classes:

(*a*) Flat Work and (*b*) Round Work. While it cannot be said that each class has its distinctive line of measuring tools, the workman who handles flat work only will usually have a somewhat different set of measuring tools from the workman on round work.

FLAT WORK

In general the worker on flat work will need to be provided with steel rules, dividers, protractors, straight

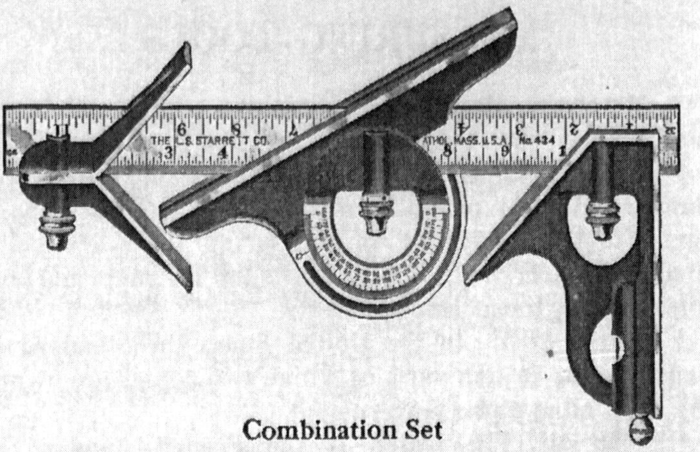

Combination Set

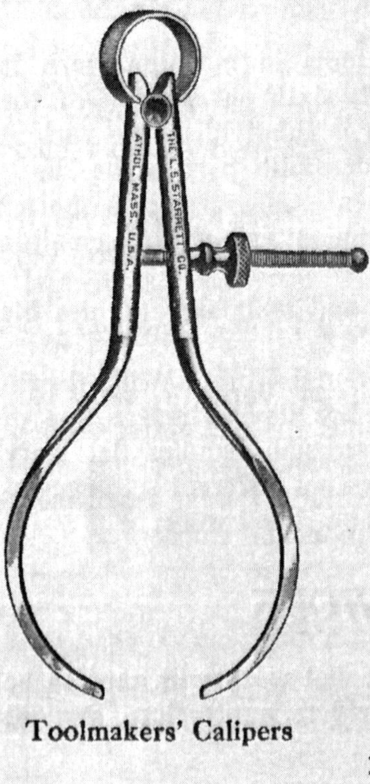

Toolmakers' Calipers

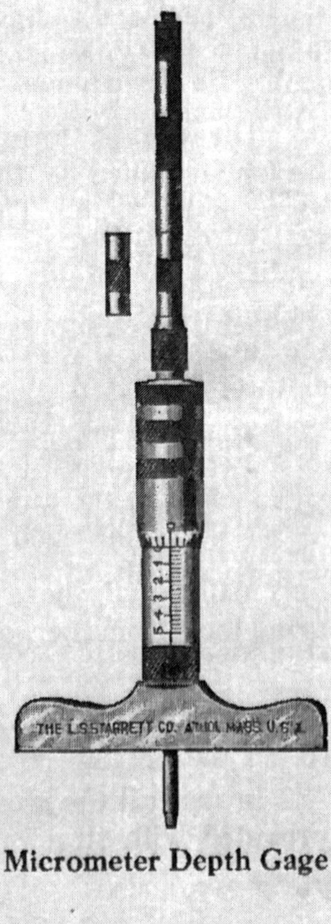

Micrometer Depth Gage

edges, steel squares, surface, height, depth, and thickness gages, center punches, parallels, slide calipers, etc.

ROUND WORK

For round work the measurements are by contact, and the usual tools are those having contact points. Contact measurements are made in two ways: (*a*) The contact tool is first set to some standard of length, as, for example, a steel rule, or a standard gage. The "set" dimension may then be used as a standard for testing the work. (*b*) The reverse of this method may be used for determining sizes, viz.: by first setting the contact points to the surfaces of the work, afterward using the steel rule or standard gage to read the size.

"FEEL"

The accuracy of all contact measurements is dependent upon the sense of touch (feel). In the case of skilled workmen, as, for example, toolmakers, the sense of touch is highly developed. Using suitable contact measuring tools, the skilled mechanic can readily "feel" the difference in contact made by changes of dimensions as small as 0.00025".

In the human hand the sense of touch is most prominent in the finger-tips. Therefore the contact measuring tool should be held by the fingers only, and in such a way as to bring it in contact with the finger-tips. If the tool is harshly grasped by the fingers, the sense of touch or feel is much reduced. For this reason the tool should be delicately and lightly held instead of gripped tightly.

The more common tools for contact measurements are inside and outside calipers, used in conjunction with steel rules, plug and ring gages, and dimension blocks.

While it is possible to transfer by "feel" a length with an error not exceeding one-quarter of one thousandth inch, the results are not always easily read; for this reason mechanics prefer to use direct reading tools for the more accurate contact work. Two methods of direct reading are in common use.

VERNIER CALIPERS

This tool is a combination of contact points and steel rules. One of the contact points is a fixed part of a graduated steel rule, while the other contact point is a part of a graduated slider mounted upon the blade of the first. By combining the use of the separate scales, direct readings of one-thousandth part of an inch are readily made.

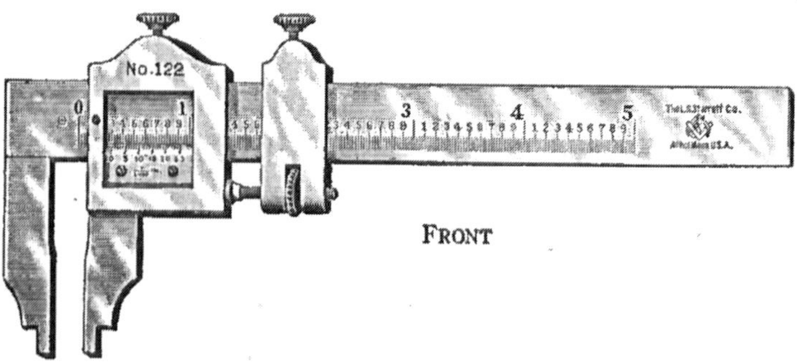

FRONT

VERNIER HEIGHT GAGE

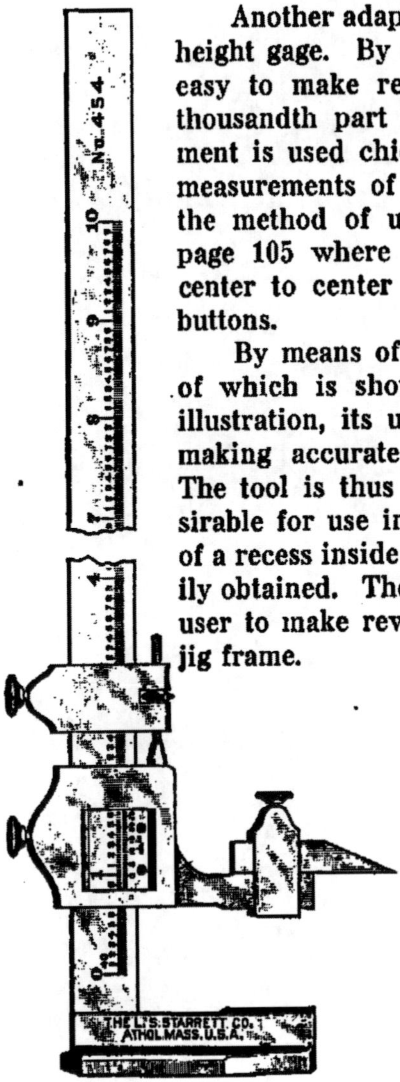

Another adaptation of the vernier is the height gage. By means of the vernier it is easy to make readings as minute as one thousandth part of an inch. This instrument is used chiefly where close, accurate measurements of height must be obtained; the method of using is clearly shown on page 105 where it is used in finding the center to center distance of a pair of jig buttons.

By means of suitable adjustments, one of which is shown on the accompanying illustration, its use is extended to include making accurate measurements of depth. The tool is thus rendered particularly desirable for use in jig-making for the depth of a recess inside the jig frame may be readily obtained. The removable jaw allows the user to make reverse measurements on the jig frame.

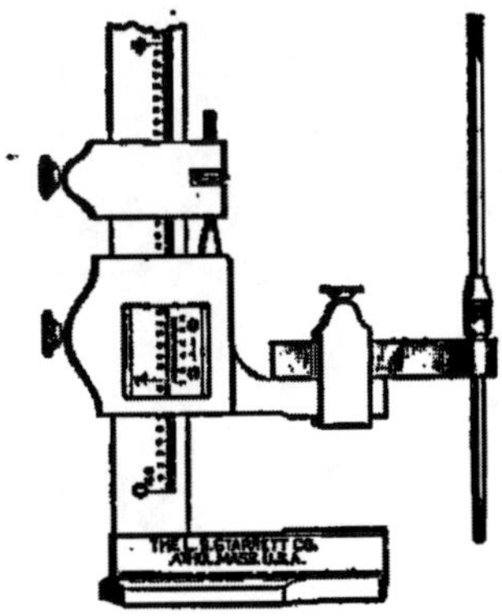

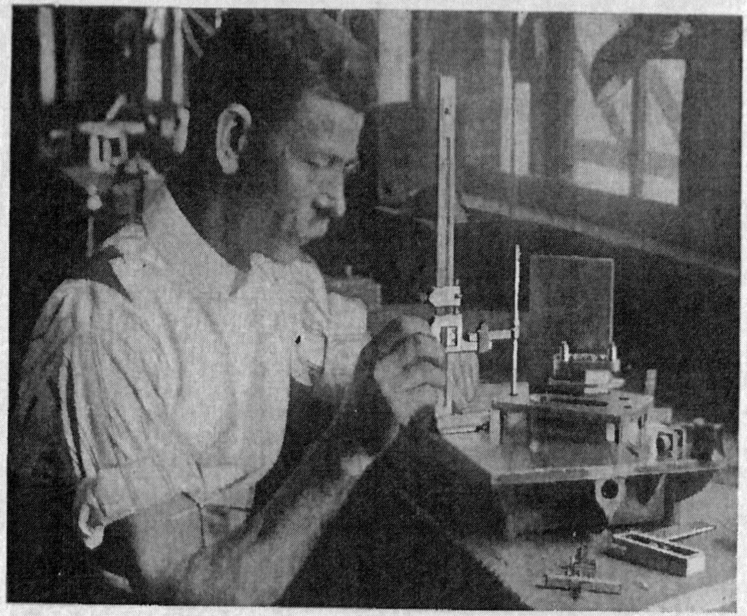

MICROMETER CALIPERS

With the invention of the micrometer screw there came into use a new method of direct readings in contact measurements. The great accuracy of the micrometer screw becomes evident when it is realized that threaded spindles with a limit of error of 0.001" in one-foot lengths are commercially possible. In micrometer construction with a used length of screw thread of one inch only, the error is negligible. A micrometer head consists of a spindle, threaded forty to the inch, fitted through a threaded sleeve, having an enclosing thimble fastened to its outer end. Suitable graduations made axially on the threaded sleeve combined with the graduations on the edge of the rotating thimble give direct readings of one-thousandth part of one inch. By means of a vernier scale used on the rear of the sleeve direct contact readings as small as one ten-thousandth part of one inch can be readily made.

Micrometer screws are mounted in a frame which may be varied in shape and size to render it convenient for the desired purposes. The contact points are also shaped to the particular use desired, and instruments of this type in a variety of styles and of the highest degree of accuracy, convenience, and finish are purchasable, for either inside or outside measurements.

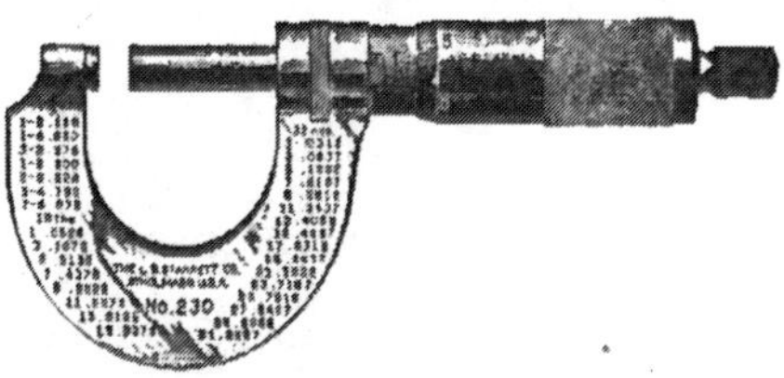

For measurement by thousandths up to one-half inch.

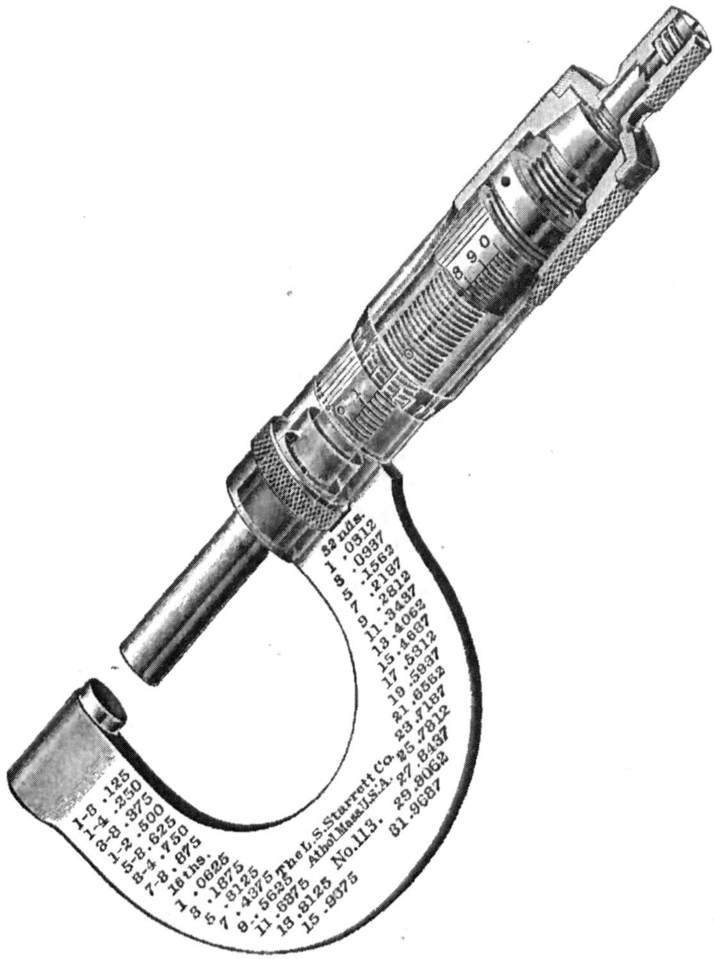

Micrometer Measurements

The limit of accuracy obtained by measuring between contacts depends on the graduations on the instrument. It is evident that as the fineness of the graduation increases, the chances for mistaking one graduation for another also increase so that some other method of determining extremely accurate measurements must be devised.

The common instrument for making such measurements is known as a micrometer-caliper. It combines the double contact of the slide calipers with a screw adjustment which may be read with great accuracy.

HOW TO READ A MICROMETER

The pitch of the screw threads on the concealed part of the spindle is forty to an inch. One complete revolution of the spindle, therefore, moves it lengthwise one fortieth (or twenty-five thousandths) of an inch. The sleeve D is marked with forty lines to the inch, corresponding to the number of threads on the spindle.

Each vertical line indicates a distance of one-fortieth of an inch. Every fourth line is made longer than the others, and is numbered 0, 1, 2, 3, etc. Each numbered

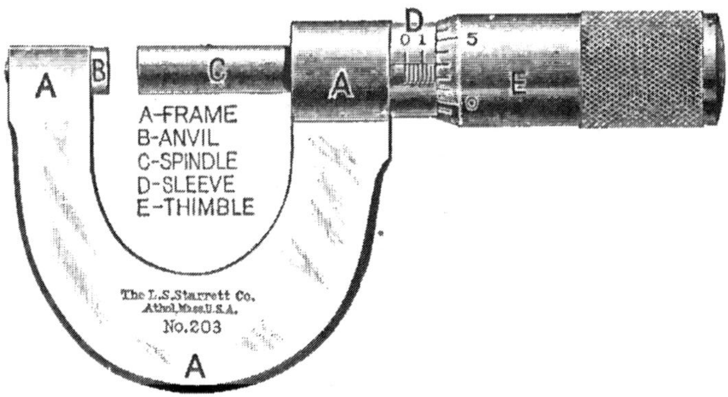

line indicates a distance of four times one-fortieth of an inch, or one tenth.

The beveled edge of the thimble is marked in twenty-five divisions, and every fifth line is numbered, from 0 to 25. Rotating the thimble from one of these marks to the next moves the spindle longitudinally one twenty-fifth of twenty-five thousandths, or one thousandth of an inch. Rotating it two divisions indicates two thousandths, etc. Twenty-five divisions will indicate a complete revolution, .025 or one-fortieth of an inch.

To read the micrometer, therefore, multiply the number of vertical divisions visible on the sleeve by twenty-five, and add the number of divisions on the bevel of the thimble, from 0 to the line which coincides with the horizontal line on the sleeve. For example, in the engraving, there are seven divisions visible on the sleeve. Multiply this number by twenty-five, and add the number of divisions shown on the bevel of the thimble, 3. The micrometer is open one hundred and seventy-eight thousandths. ($7 \times 25 = 175$ and $175 + 3 = 178$.)

HOW TO READ A VERNIER

Readings in ten thousandths of an inch on caliper squares, micrometers, etc., are obtained by the use of a Vernier, named from Pierre Vernier, who invented the device in 1631. For the Vernier caliper, the scale on the tool is graduated in fortieths of an inch (0.25). On the Vernier plate is a distance divided into twenty-five parts, and these twenty-five divisions occupy the same distance as twenty-four divisions on the scale. The difference between one of the twenty-five spaces and one of the twenty-four spaces is one twenty-fifth of one-fortieth, or one thousandth of an inch.

To read the tool, note how many inches, tenths (or .100), and fortieths (or .025) the 0 mark on the Vernier

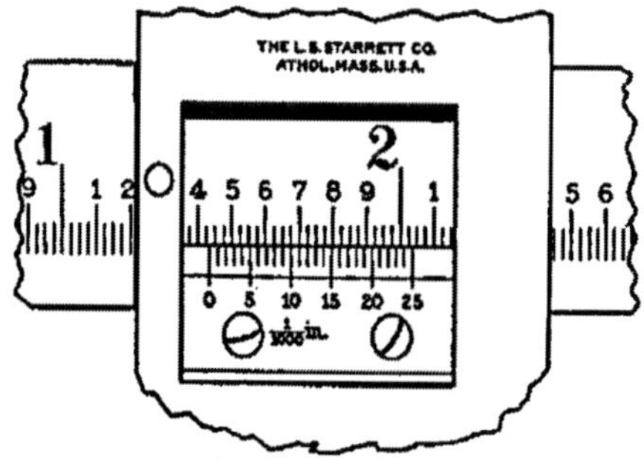

is from the 0 mark on the scale; then note the number of divisions on the Vernier from 0 to a line which exactly coincides with a line on the scale.

In the engraving above, the Vernier has been moved to the right one and four-tenths and one-fortieth inches

(1.425"), as shown on the scale, and the eleventh line on the Vernier coincides with a line on the scale. Eleven thousandths of an inch are, therefore, to be added to the reading on the scale, and the total reading is one and four hundred and thirty-six thousandths inches (1.436"), which is the distance the jaws have been opened.

HOW TO READ A VERNIER MICROMETER

Readings in ten thousandths of an inch are obtained ON THE MICROMETER by the use of a Vernier, which operates on the same principle as the Vernier on the caliper. In this case, however, ten divisions on the sleeve occupy the distance of nine divisions on the thimble. The difference between the width of one of the ten spaces and one of the nine spaces is one-tenth of a

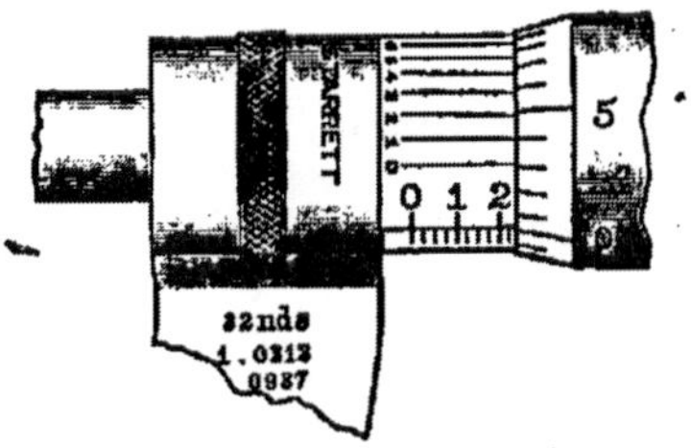

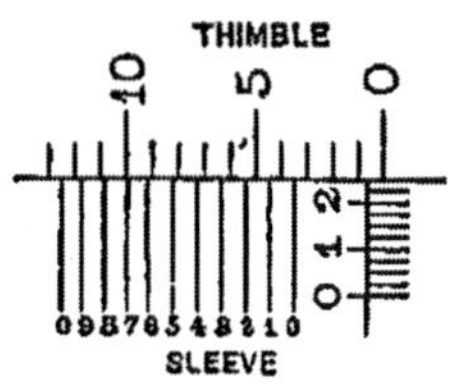

C

division on the thimble. Now each division on the thimble represents one-thousandth of an inch, and one-tenth of one-thousandth equals one ten-thousandth. To read a ten-thousandth micrometer, first note the thousandths as in the ordinary micrometer. Then observe the line on the sleeve which coincides with a line on the thimble. In the diagram shown above there are nine vertical divisions visible on the sleeve, and 9 × 25 = 225, so that the reading of the ordinary micrometer would be .225. Line marked "7" on the sleeve coincides with a line on the thimble and, therefore, we add seven to the reading of the ordinary micrometer. This seven is seven ten-thousandths (.0007), and the readings will be .2257.

Half-Inch Micrometer

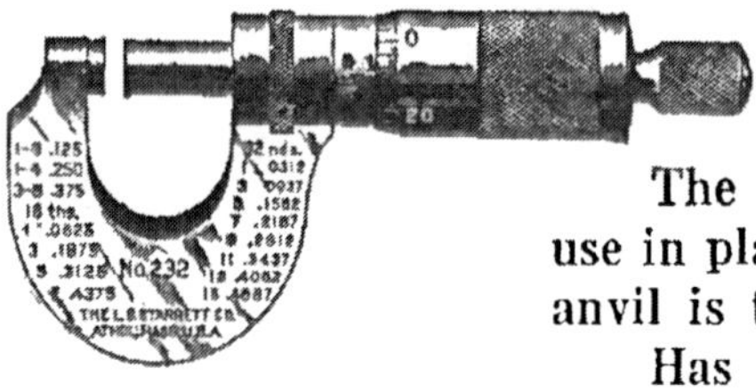

For measurement by thousandths up to one-half inch.

The anvil is shortened, for use in places where the ordinary anvil is too long to be inserted.

Has lock nut and ratchet stop.

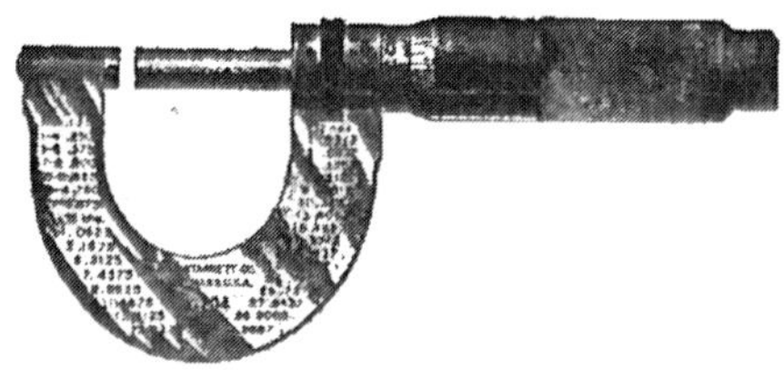

Quick-Adjusting Micrometer

Has ratchet stop and lock nut.

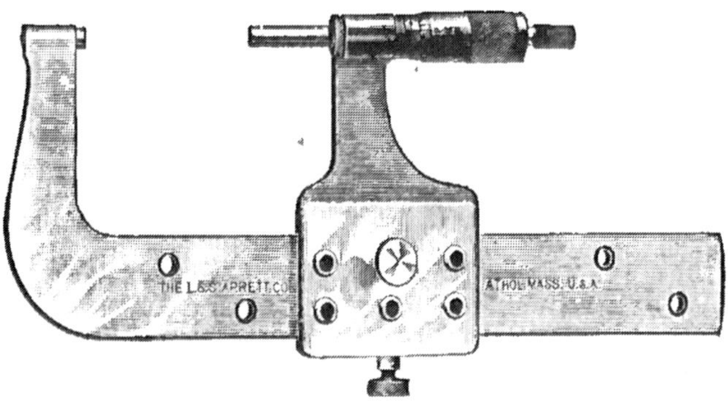

Six-Inch Micrometer

For measuring round work to 4½ inches and flat work to 6 inches.

OPERATION AND ADJUSTMENT OF MICROMETERS

QUICK MEASUREMENTS. A micrometer having the quick-adjusting feature can be instantly opened or closed to any size within its capacity. Pressure of the finger on the end of the plunger allows the spindle to move instantly to the desired size without turning the thimble. When the finger is removed, fine adjustments may be made in the usual way.

MICROMETER AS A GAGE. By means of a knurled lock nut the spindle can be firmly fixed in position, making the micrometer a solid gage. Turning the lock nut contracts a split bushing around the spindle, keeping it central and true.

READJUSTMENT FOR WEAR. When slight wear makes correction necessary, the readjustment is accomplished by various means depending upon the kind of micrometer. With the Starrett micrometer the anvil is fixed, not movable, and correction is quickly made by inserting a spanner wrench and turning until the line on the sleeve coincides with the zero on the thimble. This feature does away with the frequent use of a test piece.

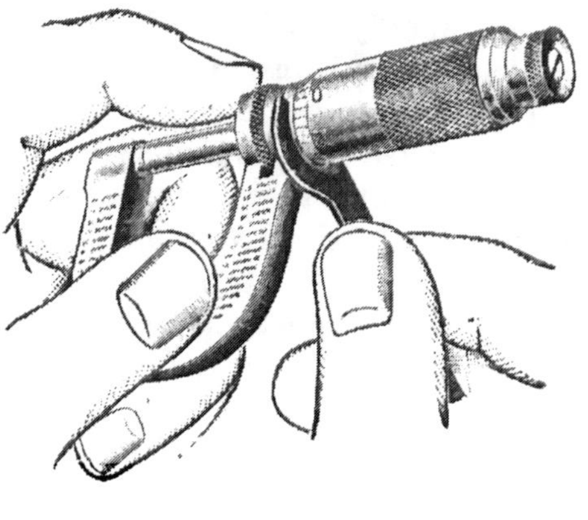

TRANSFERRING MEASUREMENTS

Transferring a measurement may be a delicate job or not, wholly depending upon the degree of accuracy sought. The most common of all machine-shop tools for transferring measurements are steel rules and spring calipers. With these tools, either in combination or used separately, are made the bulk of common machine-shop measurements, whether those of inside or outside surfaces.

STEEL RULES

These are thin blades of steel of varying lengths, widths, and thicknesses, usually graduated in inches and various subdivisions of the inch upon each edge of both sides and often at the ends. The makers term the various subdivisions of the inch by graduation numbers, for example, No. 4 Graduation, 1st. edge 64ths; 2d. edge 32ds; 3d. edge 16ths; 4th. edge 8ths. By means of sliding or fixed attachments a great variety of length measurements may be made with the ordinary steel rule.

SPRING CALIPERS

The most commonly used tool for contact measurements is the ordinary spring caliper, which is used for measuring over surfaces or between surfaces. In shop language this is called making-outside-or-inside measurements. The legs of the spring caliper are curved down, to make two opposite contact points, the distance between being controlled by a screw which works against a tension spring. For either outside or inside measurements they may be set to or they may be read to a graduated steel rule. In this way a workman can transfer lengths with an error of less than 0.002". Where specially accurate spring caliper measurements are desired, fixed gages are used for setting the contact points. The degree of accuracy of contact is dependent upon what the workman terms "feel." To accurately transfer a dimension with spring calipers the sense of "feel" must be well developed by the workman, for the contact points are at the ends of very slender arms.

Spring calipers, both for inside and outside work, can be set to dimensions either larger or smaller than the gages used by introducing thickness strips between the contact points and the over or inside surfaces.

Hard, thin tissue-paper may be used as thickness strips, or, better still, steel thickness gages or "feelers."

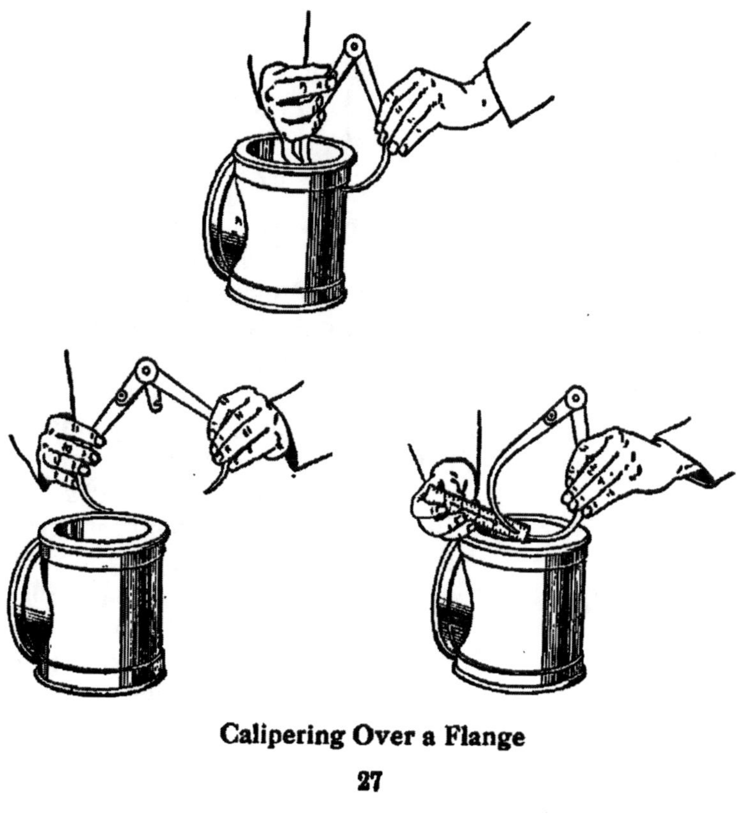

Calipering Over a Flange

SPRING DIVIDERS

In this tool the contacts are points at the ends of straight legs. Dividers are used for measuring dimensions between lines or points, for transferring lengths taken direct from a graduated steel rule, or for scribing

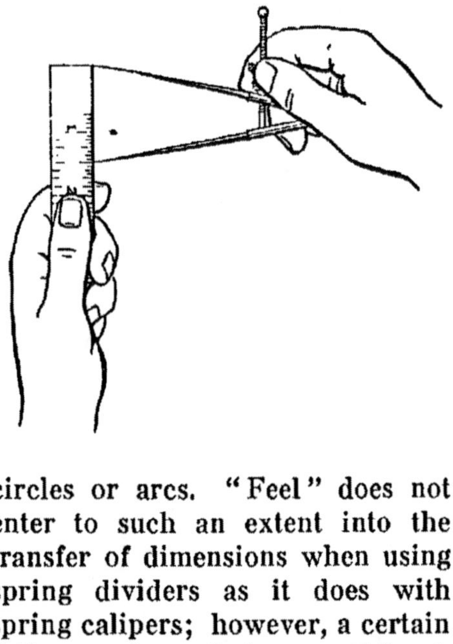

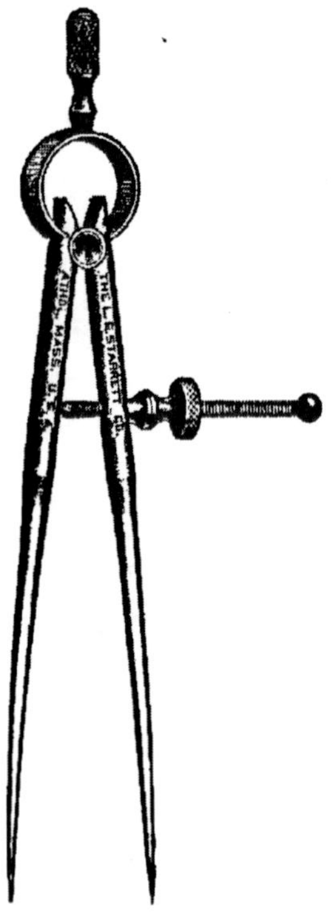

circles or arcs. "Feel" does not enter to such an extent into the transfer of dimensions when using spring dividers as it does with spring calipers; however, a certain delicacy of touch is essential. A magnifying glass is a wonderful help for the accurate transfer of dimension with dividers. If a considerable length is to be transferred, it is best to use the type where the points are adjustable along a bar, known as a Universal Divider, for the points do not then incline to the surfaces worked upon.

FITS AND FITTING

In machine construction many of the parts bear such a close and important relation to one another, that a certain amount of hand fitting is essential to make the surface contacts as they should be. If the surfaces in contact are to move on each other the fit is classed as a sliding or running fit. If the surfaces are to make contact with sufficient firmness to hold them together under ordinary use, the fit is classed either as a driving, shrink, or forced fit.

SLIDING FIT. Under this head may be classed the fitting of cross and traversing slides of lathes, milling machines, drilling machines, boring machines, grinding machines, and planers. In most of these fits the moving and stationary parts are held in contact with each other by means of adjustable contact strips or gibs, *sometimes known as packing strips.* In some cases, such as the tables of grinding and of planing machines, their weight keeps them in sufficiently close contact.

RUNNING FITS. The journal bearings of spindles, crank shafts, line shafting, etc., are classed under this heading.

FORCED FITS AND SHRINK FITS. Under this head are classed those fits where the separate parts must become in use as if they were a single piece; as, for example, the crank pins and axles in locomotive driving wheels, the cutter heads and spindles of numerous wood-working machines, as well as many other cases.

LIMITS. In the case of running and of sliding bearings a certain amount of hand fitting is necessary to obtain desired results, and in all cases certain limiting requirements obtain. In sliding and running bearings the limits are usually those of alignment and of contact, while in either journal bearings or in flat sliding bearings it is essential that certain accurate contact between the surfaces shall be made, and there will also be a limit of alignment with other parts of the machine. For example, in the engine lathe the ways or vees and the cross slide of the tool carriage must be parallel to or at right-angles to the axis of the spindles within set limits. In engine lathe construction the limit set for this is 0.001" in a foot of length. In testing the parts use is made of the Universal Test Indicator with the needle reading on a dial or upon a sector arm. The indicator may be clamped to a test bar, a straight edge, or direct to the lathe spindle; also, if desired, it can be and often is held upon a special slider stand fitted to the vees of the machine.

In the making of shrinkage and forced fits the limits are usually those of size. The amount of pressure necessary to place the two parts together is the limiting fact in the case of forced fits. In forcing the axles into locomotive driving wheels, the specifications may limit the pressure to between one hundred to one hundred and fifty tons. However specified, it in fact reduces to limits of size and the use of measuring tools. These can be of the direct reading contact type, as the micrometer and vernier bar, or of the indirect reading contact type, as, for example, the ordinary spring caliper used in conjunction with thickness gages or "feelers."

AMOUNTS TO LEAVE. Where pins, spindles, etc., are to be forced into holes, or where collars, hubs, flanges, and other machine parts are to be shrunk on to spindles, it is customary to make the diameter allowance upon the spindle rather than upon the hole. The amount which it is necessary to add to the spindle or shaft diameter must of necessity vary with the length and diameter of the hole, the metals used, and the form of the surrounding hub. The following tables give certain practice.

Allowances for Different Classes of Fits — Table 1

(Newall Engineering Co.)

Class	Tolerances in Standard Holes*						
	Nominal Diameters	Up to ½″	9/16″–1″	1 1/16″–2″	2 1/16″–3″	3 1/16″–4″	4 1/16″–5″
A	High Limit	+0.0002	+0.0005	+0.0007	+0.0010	+0.0010	+0.0010
	Low Limit	−0.0002	−0.0002	−0.0002	−0.0005	−0.0005	−0.0005
	Tolerance	0.0004	0.0007	0.0009	0.0015	0.0015	0.0015
B	High Limit	+0.0005	+0.0007	+0.0010	+0.0012	+0.0015	+0.0017
	Low Limit	−0.0005	−0.0005	−0.0005	−0.0007	−0.0007	−0.0007
	Tolerance	0.0010	0.0012	0.0015	0.0019	0.0022	0.0024
	Allowances for Forced Fits						
F	High Limit	+0.0010	+0.0020	+0.0040	+0.0060	+0.0080	+0.0100
	Low Limit	+0.0005	+0.0015	+0.0030	+0.0045	+0.0060	+0.0080
	Tolerance	0.0005	0.0005	0.0010	0.0015	0.0020	0.0020
	Allowances for Driving Fits						
D	High Limit	+0.0005	+0.0010	+0.0015	+0.0025	+0.0030	+0.0035
	Low Limit	+0.0002	+0.0007	+0.0010	+0.0015	+0.0020	+0.0025
	Tolerance	0.0003	0.0003	0.0005	0.0010	0.0010	0.0010
	Allowances for Push Fits						
P	High Limit	−0.0002	−0.0002	−0.0002	−0.0005	−0.0005	−0.0005
	Low Limit	−0.0007	−0.0007	−0.0007	−0.0010	−0.0010	−0.0010
	Tolerance	0.0005	0.0005	0.0005	0.0005	0.0005	0.0005
	Allowances for Running Fits†						
X	High Limit	−0.0010	−0.0012	−0.0017	−0.0020	−0.0025	−0.0030
	Low Limit	−0.0020	−0.0027	−0.0035	−0.0042	−0.0050	−0.0057
	Tolerance	0.0010	0.0015	0.0018	0.0022	0.0025	0.0027
Y	High Limit	−0.0007	−0.0010	−0.0012	−0.0015	−0.0020	−0.0022
	Low Limit	−0.0012	−0.0020	−0.0025	−0.0030	−0.0035	−0.0040
	Tolerance	0.0005	0.0010	0.0013	0.0015	0.0015	0.0018
Z	High Limit	−0.0005	−0.0007	−0.0007	−0.0010	−0.0010	−0.0012
	Low Limit	−0.0007	−0.0012	−0.0015	−0.0020	−0.0022	−0.0025
	Tolerance	0.0002	0.0005	0.0008	0.0010	0.0012	0.0013

* Tolerance is provided for holes, which ordinary standard reamers can produce, in two grades, Classes A and B, the selection of which is a question for the user's decision and dependent upon the quality of the work required; some prefer to use Class A as working limits and Class B as inspection limits.

† Running fits, which are the most commonly required, are divided into three grades: Class X for engine and other work where easy fits are wanted; Class Y for high speeds and good average machine work; Class Z for fine tool work.

LIMITS OF TOLERANCE

While it is possible to produce machine parts with measurements refined to any degree of accuracy, extreme precision may prove too costly for commercial work.

To avoid waste of time, labor, and money, the Taft-Peirce Manufacturing Company has formulated a set of rules which defines the degree of accuracy to be expected in those cases where specifications and drawings do not call for greater precision than the rules provide for.

(1) Full information regarding limits of tolerance should be clearly shown by drawings submitted, or be definitely covered by written specifications to which reference must be made by notations on the drawings.

(2) Where the customer fails to supply proper data as to limits, this Company's Engineers will use their best judgment in deciding just what limits it may be advisable to work to. The Company will not, in any event, assume responsibility for possible excessive cost brought about through working to closer limits than may be necessary nor for permitting greater latitude than may subsequently be found to be proper.

(3) Where dimensions are stated in vulgar fractions with no limits of tolerance specified, it will be assumed that a considerable margin for variation from figured dimensions is available; unless otherwise ordered, the Company's Engineers will proceed according to the dictates of their best judgment as to what limits should be taken.

(4) For all important dimensions *Decimal* figures should be used and limits clearly stated on detail drawings. If *Decimal* figures are not used for such dimensions a notation referring to the degree of accuracy required must be placed prominently on the drawing.

(5) It is frequently necessary to reduce fractions

representing fourths, eighths, sixteenths, thirty-seconds, and sixty-fourths to decimal equivalents. When a dimension of this character is expressed in a decimal equivalent and carried out to three, four, or five places and limits are not specified it will be assumed that a limit of plus or minus .0015 is permissible unless otherwise ordered.

(6) Where dimensions are stated in decimal figures derived by other processes than those explained in paragraph five, but with limits not specified, the following variations from dimensions stated may be expected:

Two place decimals		.005	plus or	minus
Three	" "	.0015	"	"
Four	" "	.0005	"	"
Five	" "	.0002	"	"

(7) Where close dimensions, such as the location of holes from center to center in jigs, fixtures, machine parts, and other exact work of like character are required, detail drawings should be prominently marked "ACCURATE" and clear instructions be given.

(8) The dimensions of internal cylindrical gages, external ring gages, snap gages, and similar work specified to be hardened, ground, and lapped, will be obtained as accurately as the best mechanical practice applying to commercial work of the particular grade specified will permit.

(9) As drilled holes vary in size from .002" to .015" (and in some cases even more) over the size of the drill used, those which require to be made accurately to definitely specified sizes should be either reamed, ground, or lapped, and detail drawings thereof should bear notations accordingly.

(10) U. S. Standard form of thread and pitches will be used for ¼-inch and all sizes above. A. S. M. E. Standard will be used for numbered sizes below ¼-inch. In the absence of specifications to the contrary, U. S. Standard form of thread will be used for all SPECIAL sizes.

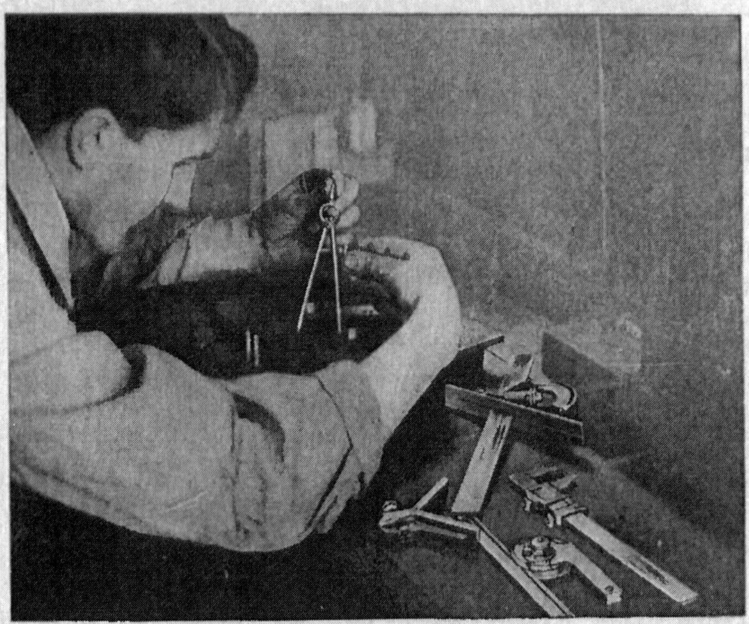

BENCH WORK

Bench work includes laying out, chipping, filing, polishing, hand reaming, hand tapping, and all the many shop jobs done at the bench or in a vise.

LAYING OUT. This is the shop term which includes the placing of lines, circles, and centers upon curved or flat surfaces for the guidance of the workman. It is somewhat analogous to mechanical drawing. It differs in one important respect, however, that while a line drawing is seldom scaled and therefore exact accuracy of spacing is not required; in laid out work, the lines, circles, centers, etc., are to be followed exactly. All lines, centers, etc., should therefore be exactly located and placed, and all scriber, divider, and center points should, while in use, be exact and sharp. Particular care must be maintained to insure fine and accurate laying out.

PREPARING THE SURFACE. If work of no special accuracy is desired, carefully rubbing chalk, or white lead mixed with turpentine, upon the surface of the work will be sufficient as a coating. For fine exact layouts a special marking solution must be used. The one in common shop use is a mixture of one ounce copper sulphate to four ounces water. A little nitric acid may with advantage be added. This solution applied to a cleaned iron or steel surface gives a dull coppered surface, and the finest line scribed upon it is brilliantly visible.

SCRIBING LINES. The usual scribing points are those common to dividers, hermaphrodite calipers, scratch awls, scratch gages, surface gages, and trammel points. Combined with the scribing points, may be used steel rules, bevel protractors, steel squares, steel straight edges, levels, end measuring rods, micrometer or vernier height and depth gages, and the various center punches. Ability to so combine and make use of the various tools

as to insure accuracy is a considerable asset to the laying-out man.

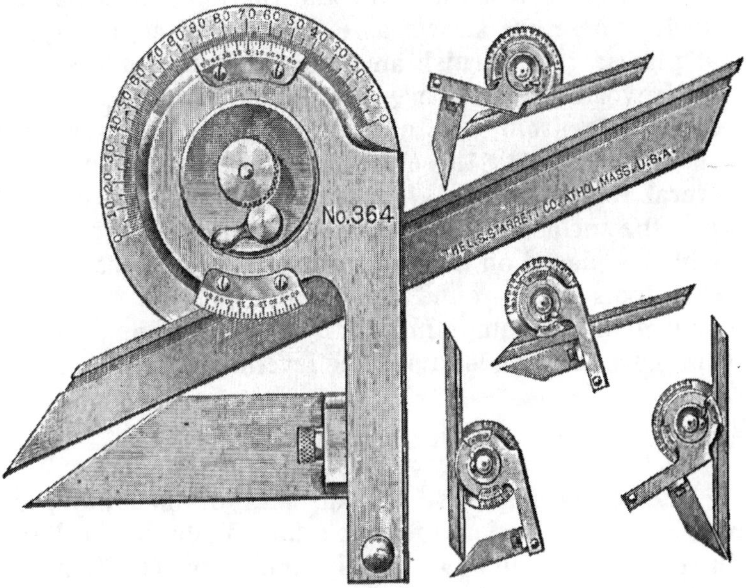

PROTRACTORS

As made for machine-shop use the common protractor is provided with attached straight edges, and can be used either to measure or to lay off lines at an angle to each other. Measuring the angularity of two or more lines with a protractor is termed "reading the angles." As oftentimes its use is determining the angle made by two surfaces (a bevel), the tool is usually termed a bevel protractor. Protractors for common shop use are graduated to degrees through a length of circumference of one hundred and eighty degrees. An attached vernier enables the user to read angles to one-twelfth of a degree (five minutes).

LAYING OUT PLATE. If desirable results are to be

obtained in laying out flat work, special metal plates upon which to rest the work and the tools must be provided. These are known as leveling, surface, or laying-out plates; they furnish an accurate plane surface upon which work and tools may be placed. The size of these plates varies from those of small areas used in laying out small jigs, etc., to those for large pieces, having sides several feet in length. The work may be laid directly upon the surface of the plate or held upon leveling strips or blocks placed on the plate, and the gages, squares, and other tools used around the work. In other cases it is convenient to clamp the work to knee or angle irons, which are then placed upon the leveling plate.

CHIPPING

Formerly many of the surfaces of machine parts were hand-chipped and filed to a fit. While the mechanic in the modern shop can usually find methods of machining most of the surfaces he needs to fit up, there are still occasions when the work has to be hand-chipped.

TOOLS USED. The common chipping tools are a hand hammer and a hand chisel. The hand hammer should weigh not less than three-quarters of a pound nor over two pounds, and may be either of the ball peen or flat peen type. A chipping hammer should balance well in the hand when fitted to a handle not more than sixteen inches long. The handle near where it enters the hammer should be thinned and worked down to a shank that is somewhat flexible, so that the shock to the arm and hand will be less. The face of a good chipping hammer should crown slightly.

Chipping chisels, ordinarily termed cold chisels, are of various sorts, and are often known by the shape of the cutting end; for example, flat, cape, roundnose, diamond, and gouge chisels. The steel from which they are made should be eighty to ninety point carbon, of octagon cross-section, with the cutting end forged to the desired shape, well packed by the forge hammer, hardened, and the temper drawn to a medium blue. The

hammer end of the chisel should be forged from the octagon to a reduced round but not hardened. Flat-chipping and cape chisels should be ground with straight, symmetrical, cutting edges, at as acute an angle as the nature of the work will permit.

In hand chipping the hammer handle should be grasped near the end and the hammer swung free from over the shoulder with an easy forearm movement. Hold the chisel loosely in the hand at an angle with the work that permits an even chip of right depth. The vision should be directed to the cutting edge of the chisel, rather than at the end struck by the hammer. Avoid gripping hammer or chisel tightly, as this rapidly tires the hand and arm.

In shops which have compressed air, use is made of the modern pneumatic chipping hammer, which does remarkable work of the heavier sorts.

FILING

The file is essentially a finishing tool, and in skilled hands surfaces may be made very accurate and smooth.

Files are designated thus (*a*) by their length — this does not include the tang; (*b*) by their cross-section, as, for example, square, round, half-round, triangular, flat, knife-edge, etc.; (*c*) by their cut — single or double cut; (*d*) by the degree of coarseness.

Files for some purposes are made tapered in their length, and for other uses have straight sides. The degrees of coarseness are designated by the following names as rough, coarse, bastard; 2d — cut, smooth, and dead smooth; extra fine files are designated by numbers, No. 00, No. 0, No. 1, etc., to No. 8. The degree of coarseness varies with the length, for example, an 8-inch file second cut is coarser than a shorter file bastard cut. This confuses the user somewhat, unless he is familiar with practice.

Single-cut files are those having teeth made by single parallel cuts across the face at an angle of twenty-five degrees. In double-cut files the teeth are made by breaking up the single cuts into points by a second cut made at an angle with the first.

Rasp files are those having teeth made by a punch. Used for hoofs, wood, etc.

HEIGHT OF WORK. This must of necessity vary with the height of the worker. A common rule is to have it the height of the worker's elbow as he stands erect. For very light free-hand filing the work may be much higher, in some cases the height of the shoulders.

POSITION OF THE HANDS. If the worker wishes to avoid tiring, position is very important; position also has direct bearing upon the quality and quantity of the product. The worker should clasp the file handle with the extended thumb on top, grasping the point with the fingers and thumb of the remaining hand with thumb on top. In heavy filing the point of the file may be grasped by the fingers and the palm of the hand with the palm on top.

In hand-filing the worker should train his hands, arms, and body to carry the file across the work with regular, even, and controlled strokes. As the file is in no sense self-guided the worker must train his body to regular controlled motions if he is to do effective work.

DRAW FILING. Used to set the grain somewhat smoother than regular cross-filing. The worker should clasp the blade of file near its ends in each hand and then draw the file, held crosswise, along the length of the work. A fine grain surface results.

TESTING FLAT FILING. Flat work is tested by the use of steel straight edges, steel squares, bevel protractors, etc.

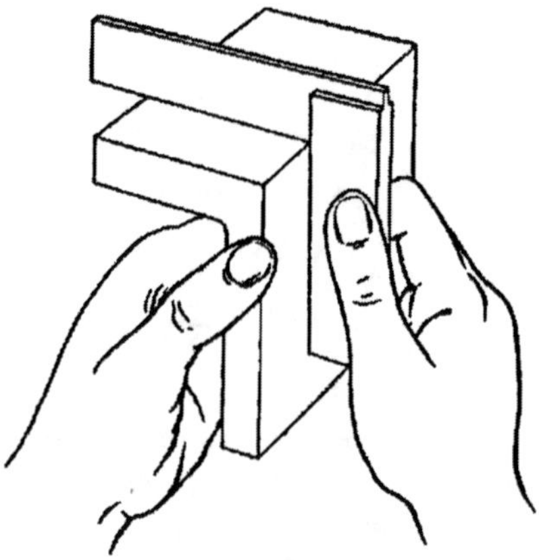

POLISHING

Where a particularly smooth surface is necessary, as, for example, journal bearings, or where brilliancy of finish is desired, the surfaces are polished with some fine abrasive. For ordinary polishing of machine parts, journals, etc., common grain abrasive is used, — glued to cloth or leather.

Grain abrasives are known by numbers, as, for example, No. 100, which means that the particles are of a size to readily pass through a sieve having one hundred meshes to the linear inch. The finer sizes are often known as flours.

GRADES OF EMERY

The numbers representing the grades of emery run from 8 to 120, and the degree of smoothness of surface they leave may be compared to that left by files as follows:

8 and 10	represent	the	cut	of	a wood rasp.
16 ,, 20	,,	,,	,,	,,	a coarse rough file.
24 ,, 30	,,	,,	,,	,,	an ordinary rough file.
36 ,, 40	,,	,,	,,	,,	a bastard file.
46 ,, 60	,,	,,	,,	,,	a second cut-file.
70 ,, 80	,,	,,	,,	,,	a smooth file.
90 ,, 100	,,	,,	,,	,,	a superfine file.
120F and FF	,,	,,	,,	,,	a dead-smooth file.

SEVERING METAL WITH HACK SAWS

Hack saws are narrow, thin blades of hardened steel with teeth cut along one edge, and are used for severing metal. They are held in suitable hand or power frames, which have the necessary adjustments for holding the blade in stiff tension. It is obvious that it requires care and good sense in using a hack-saw blade if good results are expected.

If the stock to be cut is both hard and thin, particular care is required to avoid injuring the blade.

CUTTING SPEED. When back sawing, under average conditions and without a lubricant, a cutting speed of fifty to sixty strokes per minute should be maintained. If the saw is used in a power machine, and the material is soft steel, a cutting speed of one hundred strokes per minute may be made, using a suitable lubricant. Unannealed tool steel should be cut under the above conditions at not to exceed sixty strokes per minute.

MOUNTING THE BLADE. The blade when mounted in a hand-frame should have the cutting-teeth rake for-

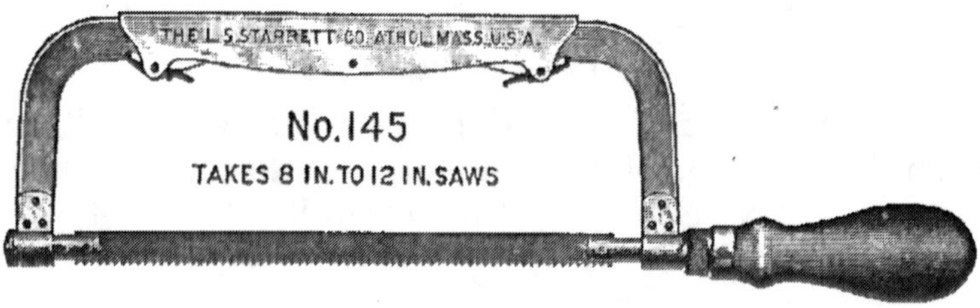

ward; that is to say, the saw should cut on the forward stroke. In machine cutting this is usually so, but not so with some makes of machines. The cutting stroke is always the pressure stroke, and the return stroke is made as light as convenient without actually lifting the blade from its work.

The blade should be under considerable tension when in use. It must be held in the plane being cut, and all tendency to bending the blade avoided. Suitable blades and frames may be purchased for almost every service, and the user should consider this fact if commercially economical results are desired.

HACK SAW MACHINE

Hack saw blades used in cutting up bar stock or structural shapes are much more efficient in a machine so designed that its several motions and adjustments can be properly controlled. Such a machine is as sensitive to the operator as a hand frame.

The machine shown above has been especially designed to efficiently operate hack saw blades. The base column carries the working parts and the work-holding vise. By means of suitable weights, the cutting pressure upon the blade may be regulated according to the material being severed, and the stroke length of the blade-carrying frame can be adjusted to use the entire blade length, no matter what diameter of bar is being severed, thus getting the full efficient service from each blade.

To avoid blade breakage through careless handling, a safety device in the form of a dash pot is connected with the blade-carrying frame to prevent the blade from being dropped suddenly upon the work. The blade-carrying frame is raised by a foot lever leaving the hands free for work adjustments and measurements. The cutting lubricant is conveyed to the blade from a tank in the column by means of a small rotary pump.

What Hack Saw to Use

No. 103 in hand frames, to cut cast steel, cast iron, tool steels and all solid metals.

No. 103B in hand frames, to cut cold rolled stock and soft metals.

No. 102 in hand frames, to cut sheet metal and tubing 16 to 18 gage.

No. 253 in hand frames, to cut sheets and tubing thinner than 18 gage.

No. 112 for heavy hand frame work and light power machines, on tool steels.

No. 112B for light power machine work on soft steel, and heavy hand frame work.

No. 114 for general work in medium weight power machines.

No. 115 on electrical conduit, pipe, brass stock, light angle and channel iron.

No. 255 on high speed machines cutting tool steels.

No. 255B on high speed machines cutting machinery steel, cast iron, etc.

No. 262 for cutting angle iron, brass stock and ornamental iron work.

No. 254 for heavy high speed machines, to cut tool steel.

No. 254B for heavy high speed machines, to cut cold rolled shafting and machinery steel.

No. 259 for cutting iron pipe, light structural iron, auto frames, etc.

No. 256 for extra heavy power machines, to cut tool steel.

No. 256B for extra heavy power machines.

DRILLING

DRILLS. A drill is an end-cutting tool, consisting usually of two cutting edges set at an angle with the axis. The more common types of drills are flat — flat-twisted — straight-fluted — spiral-fluted — and gun-barrel. The most common, and for most purposes the most efficient, type is the spiral-fluted, known as a twist drill.

Twist drills are made with two, three, or four cutting lips. The two-lip drill is used when drilling solid stock. The three and four lip drills are used for enlarging holes previously cored or drilled. When drilling solid stock with a two-lipped drill, the point of the drill controls the cutting edges, and if the drill is correctly ground the resulting hole will be reasonably round, straight, and the size of the drill. When a drill is used for enlarging holes already made, either by coring or by previous drilling, the drill is guided by its sides and a three or four fluted drill will give better results.

FORM OF POINT. In the types referred to all except gun-barrel drills are cone-pointed on the cutting end. The gun-barrel drill, used when especially straight, round, and true holes are essential, has a blunt end with a single cutting lip.

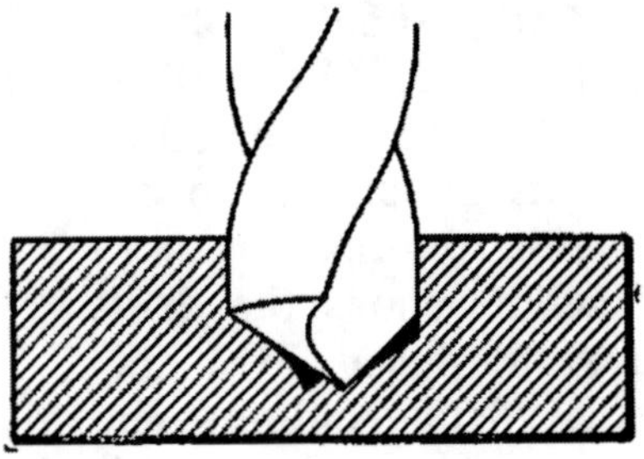

FIG. 1

A cone-pointed drill of two or more cutting lips depends for its efficient working upon four factors:

(*a*) All the cutting lips shall have the same inclination to the axis of the drill.

(*b*) Cutting lips should be of exactly equal length.

(*c*) A proper lip clearance of the surface back of the cutting edges.

(*d*) A correct angle of lip clearance.

Figs. 1, 2, and 3 show the result of careless free-hand grinding. Figs. 4 and 5 show how to test the length of the cutting lips, also their inclination to the axis.

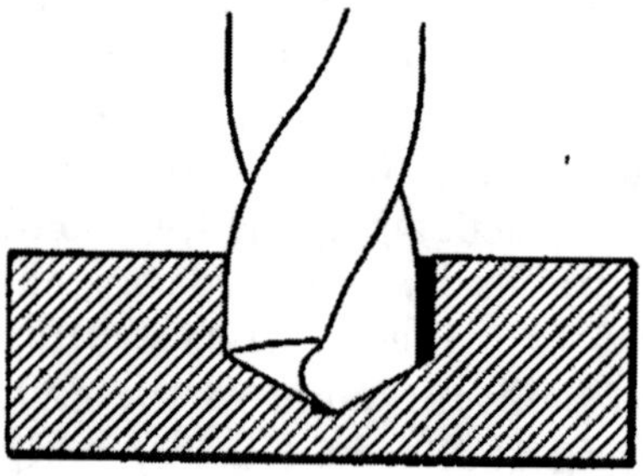

FIG. 2

After sharpening a drill free-hand, use the hand-feed at first and observe (*a*) the chips made by the cutting; (*b*) the size of the hole. If the cutting lips are shaped to a proper clearance, the chips will curl as they start from the cutting edge; but if the cutting lips lack a proper clearance the resulting chips have the appearance of being ground off rather than freely cut. If the cutting lips are of uneven length the hole will be enlarged over the diameter of the drill. Drillings from cast iron should look as in Fig. 6, and those from steel as in Fig. 7, if the drill is properly sharpened.

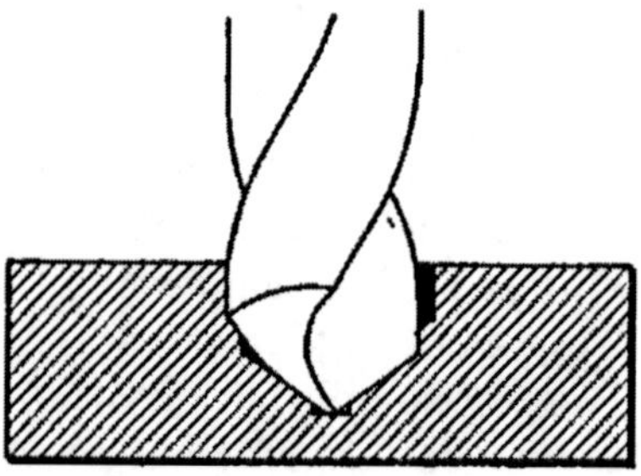

FIG. 3

Free-hand grinding results are usually so disappointing that in most machine shops the drills are sharpened in a special drill-grinding machine. The design of this machine is such, that when it is set for grinding any size of drill the cutting lips are made of equal length and of the correct form. Fig. 8 shows how the cutting lip is located to correctly grind the edges.

FEEDING THE DRILL. To get the best results from drills and drilling machines, the drill should advance into the work a definitely regulated amount for each revolution. The distance which the drill advances per revolution is termed the FEED, and must be adjusted to suit the conditions under which the work is being performed. Table No. 2 gives the feeds per revolution recommended by one manufacturer of drills. They are recommended for average conditions; they can be greatly exceeded under some conditions, but must be reduced for others.

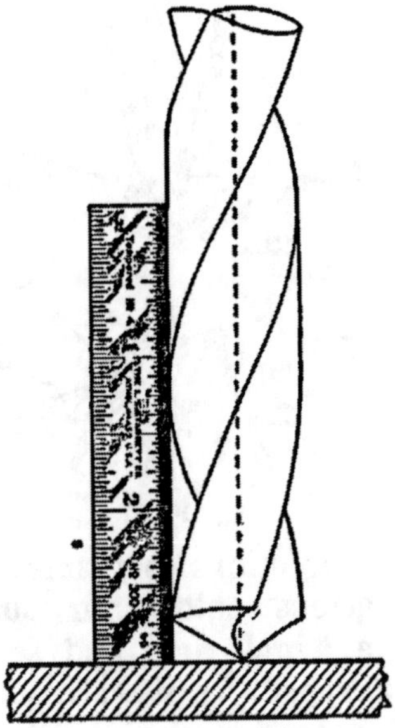

FIG. 4

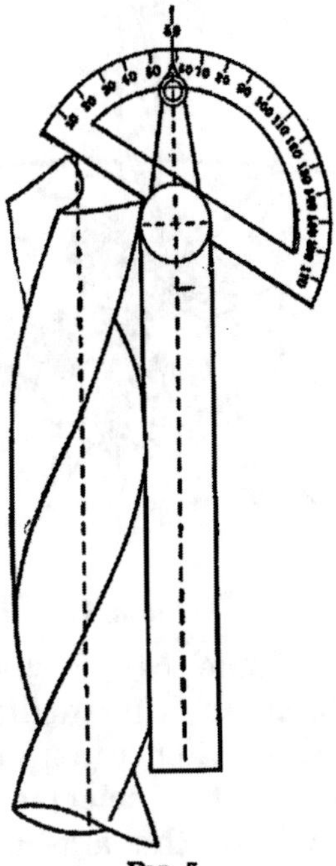

FIG. 5

Feeding the drill freehand, if skilfully done, may answer in certain cases, but is less effective than power feeds, except for small wire drills.

DRILL SPEED. This is the surface or peripheral speed of the drill in feet per minute, and is rated at the outer diameter. Under average conditions the peripheral speed recommended for carbon steel drills is thirty feet to forty feet, and for high-speed drills seventy feet to one hundred feet. Working conditions may at times cause a change in these figures. When the extreme outer corners of the cutting edges wear rapidly it is evidence of too high a surface speed.

FIG. 6

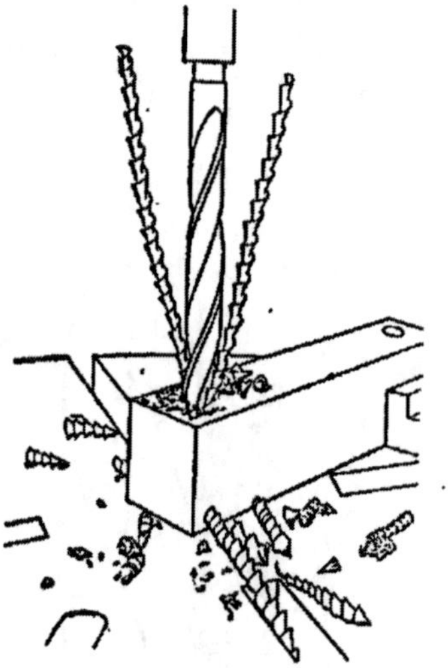

FIG. 7

Table No. 3 gives the revolutions per minute at which to run drills for various cutting or surface speeds. For example, with a 1-inch drill and seventy feet as the selected cutting speed, read across from 1-inch in the left-hand column and under heading 70′ find 267, the revolutions per minute.

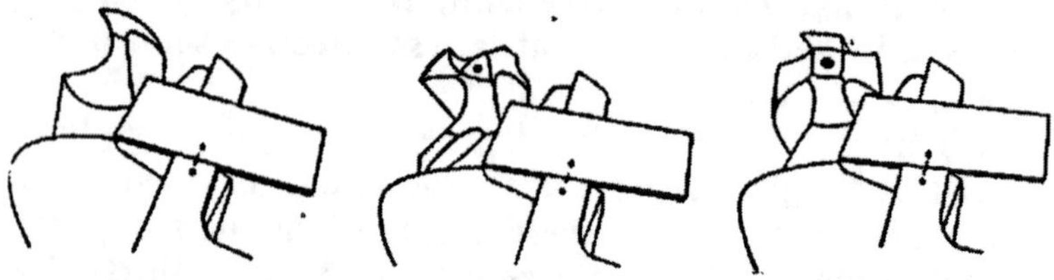

FIG. 8

Speeds and Feeds for Drilling*—Table 2

High-Speed Steel Drills

Size of Drill	Feed per Rev.	Bronze, Brass, 300 Feet	Cast Iron, Annealed, 170 Feet	Cast Iron, Hard, 80 Feet	Mild Steel, 120 Feet	Drop Forg., 60 Feet	Mal. Iron, 90 Feet	Tool Steel, 60 Feet	Cast Steel, 40 Feet
Inches	Inches	R. P. M.	R. P. M.	R. P. M.	R. P. M.	R. P. M.	R. P. M.	R. P. M.	R. P. M.
1/16	0.003	18300	10370	4880	7320	3660	3490	3660	2440
1/8	0.004	9150	5185	2440	3660	1830	2745	1830	1220
3/16	0.005	6100	3456	1626	2440	1210	1830	1220	807
1/4	0.006	4575	2593	1220	1830	915	1375	915	610
5/16	0.007	3660	2074	976	1464	732	1138	732	490
3/8	0.008	3050	1728	813	1220	610	915	610	407
7/16	0.009	2614	1482	698	1046	522	784	522	348
1/2	0.010	2287	1296	610	915	458	636	458	305
5/8	0.011	1830	1037	488	732	366	569	366	245
3/4	0.012	1525	864	407	610	305	458	305	203
7/8	0.013	1307	741	349	523	261	392	261	174
1	0.014	1143	648	305	458	229	349	229	153
1 1/4	0.016	915	519	244	366	183	275	183	122
1 1/2	0.016	762	432	204	305	153	212	153	102
1 3/4	0.016	654	371	175	262	131	196	131	87
2	0.016	571	323	153	229	115	172	115	77

Carbon Steel Drills

Size of Drill	Feed per Rev.	Bronze, Brass, 150 Feet	Cast Iron, Annealed, 85 Feet	Cast Iron, Hard, 40 Feet	Mild Steel, 60 Feet	Drop Forg., 30 Feet	Mal. Iron, 45 Feet	Tool Steel, 30 Feet	Cast Steel, 20 Feet
Inches	Inches	R. P. M.	R. P. M.	R. P. M.	R. P. M.	R. P. M.	R. P. M.	R. P. M.	R. P. M.
1/16	0.003	9150	5185	2440	3660	1830	2745	1830	1220
1/8	0.004	4575	2593	1220	1840	915	1375	915	610
3/16	0.005	3050	1728	813	1220	610	915	610	407
1/4	0.006	2287	1296	610	915	458	636	458	305
5/16	0.007	1830	1037	488	732	366	569	366	245
3/8	0.008	1525	864	407	610	305	458	305	203
7/16	0.009	1307	741	349	523	261	392	261	174
1/2	0.010	1143	648	305	458	229	343	229	153
5/8	0.011	915	519	244	366	183	275	183	122
3/4	0.012	762	432	204	305	153	212	153	102
7/8	0.013	654	371	175	262	131	196	131	87
1	0.014	571	323	153	229	115	172	115	77
1 1/4	0.016	458	260	122	183	92	138	92	61
1 1/2	0.016	381	216	102	153	77	106	77	51
1 3/4	0.016	327	186	88	131	66	98	66	44
2	0.016	286	162	77	115	58	86	58	39

The Speed of Drills—Table 3

A feed per revolution of .004 to .007 for drills 1/4 inch and smaller, and from .007 to .015 for larger is about all that should be required.

This feed is based on a peripheral speed of a drill equal to:

30 feet per minute for steel; 35 feet per minute for iron; 60 feet per minute for brass.

It may also be found advisable to vary the speed somewhat according as the material to be drilled is more or less refractory.

We believe that these speeds should not be exceeded under ordinary circumstances.

Table of Cutting Speeds

Ft. per Minute	15′	20′	25′	30′	35′	40′	45′	50′	60′	70′	80′
Diam.	REVOLUTIONS PER MINUTE										
1/16 in.	917.	1223.	1528.	1834.	2140.	2445.	2751.	3057.	3668.	4280.	4891.
1/8	459.	611.	764.	917.	1070.	1222.	1375.	1528.	1834.	2139.	2445.
3/16	306.	408.	509.	611.	713.	815.	917.	1019.	1222.	1426.	1630.
1/4	229.	306.	382.	458.	535.	611.	688.	764.	917.	1070.	1222.
5/16	183.	245.	306.	367.	428.	489.	550.	611.	733.	856.	978.
3/8	153.	204.	255.	306.	357.	408.	458.	509.	611.	713.	815.
7/16	131.	175.	218.	262.	306.	349.	393.	437.	524.	611.	699.
1/2	115.	153.	191.	229.	268.	306.	344.	382.	459.	535.	611.
5/8	91.8	123.	153.	184.	214.	245.	276.	306.	367.	428.	489.
3/4	76.3	102.	127.	153.	178.	203.	229.	254.	306.	357.	408.
7/8	65.5	87.3	109.	131.	153.	175.	196.	219.	262.	306.	349.
1	57.3	76.4	95.5	115.	134.	153.	172.	191.	229.	267.	306.
1 1/8	51.0	68.0	85.0	102.	119.	136.	153.	170.	204.	238.	272.
1 1/4	45.8	61.2	76.3	91.8	107.	123.	137.	153.	183.	214.	245.
1 3/8	41.7	55.6	69.5	83.3	97.2	111.	125.	139.	167.	195.	222.
1 1/2	38.2	50.8	63.7	76.3	89.2	102.	115.	127.	153.	178.	204.
1 5/8	35.0	47.0	58.8	70.5	82.2	93.9	106.	117.	141.	165.	188.
1 3/4	32.7	43.6	54.5	65.5	76.4	87.3	98.2	109.	131.	153.	175.
1 7/8	30.6	40.7	50.9	61.1	71.3	81.5	91.9	102.	122.	143.	163.
2	28.7	38.2	47.8	57.3	66.9	76.4	86.0	95.5	115.	134.	153.
2 1/4	25.4	34.0	42.4	51.0	59.4	68.0	76.2	85.0	102.	119.	136.
2 1/2	22.9	30.6	38.2	45.8	53.5	61.2	68.8	76.3	91.7	107.	122.
2 3/4	20.8	27.8	34.7	41.7	48.6	55.6	62.5	69.5	83.4	97.2	111.
3	19.1	25.5	31.8	38.2	44.6	51.0	57.3	63.7	76.4	89.1	102.

CUTTING COMPOUNDS. To maintain high cutting speeds, it is necessary to use a lubricant. Those recommended have stood the test of service:

For hard and refractory steel, turpentine, kerosene, or soda water.

For soft steel and wrought iron, lard oil, or soda water.

For brass, paraffine oil.

For aluminum, turpentine, kerosene, or soda water.

For cast iron, a jet of air if anything is used — usually worked dry.

LAYING OUT. Locating the centers for drilled holes upon the body of the work is termed "laying out." On the smaller jobs, laying out and drilling are usually done by the workman. Larger amounts of work warrant a skilled "layer out."

Laying out for drilling comes under two heads, viz.: APPROXIMATE and ACCURATE. Unless the holes when drilled are to match up with other holes or with fixed studs, it is enough if the center is laid off with a chalk pencil and a steel rule. For jig, tool, and experimental work, the centers must be accurately laid out and scribed upon the surface of the work. The practice is to scribe two or more lines which intersect at the exact desired point as shown in Fig. 9. Assume that the link is to

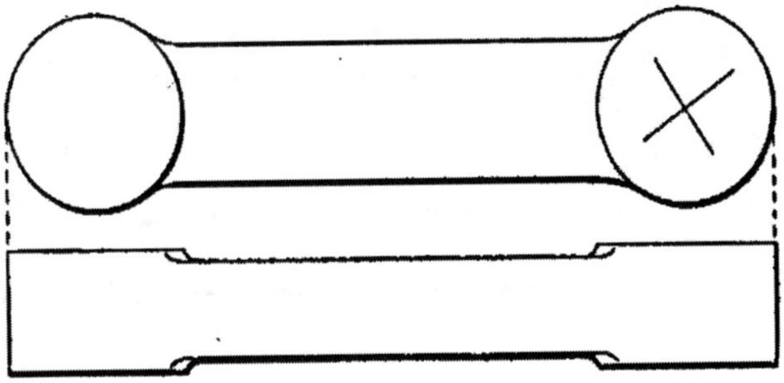

FIG. 9

connect two studs. Proceed to scribe two intersecting lines upon one of the hubs, as shown in Fig. 9, using a combination square fitted with a center head. At the intersection accurately place a light center-punch indentation. Place one leg of a spring divider with its point in the center mark and adjust the other leg to have its point touch the edge line of the hub and note the concentricity of the center. If correct, close dividers to scribe a circle the diameter of the required drilled hole, setting the points by the scale graduations upon a steel

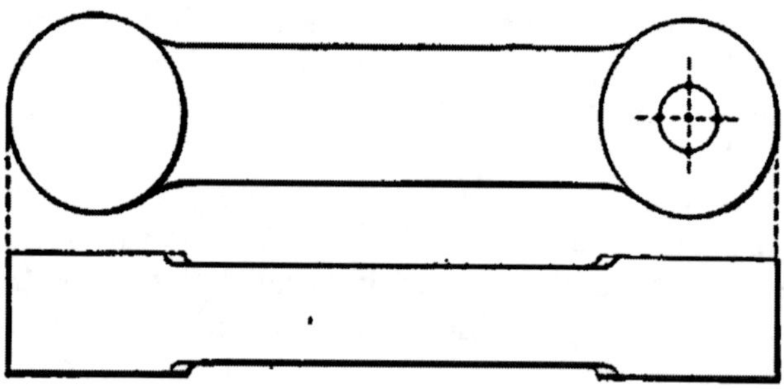

FIG. 10

rule. Locate light center-punch marks on the scribed circle as shown in Fig. 10.

When the work is laid out by another than the

FIG. 11

driller, a second circle, having a slightly greater diameter, should be scribed. This check will show whether the hole was drilled to the original lay out. If no importance is attached to the center to center distance of the holes proceed as before with the second hub. Where the center to center distance is important, set the points of the universal dividers to the center length, and with the point A, Fig. 11, in the previously located center mark scribe on the opposite hub. Scribe a short line across its face afterward, proceeding as before.

For all accurate work use the automatic center-punch, Fig. 12, and for heavy work the machinists' center-punch, shown in Fig. 13.

PREPARING THE SURFACE. For accurate laying out, clean the machined surfaces and wet the portion to be worked upon with the copper sulphate (blue vitriol) solution. When dry, the surface will distinctly show any lines which are made upon it. Chalk well rubbed into the surface is sufficient for the less accurate jobs.

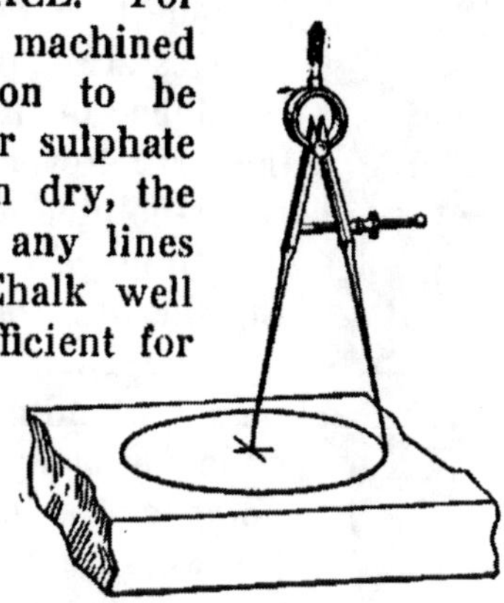

SCRIBING CIRCLES WITH DIVIDERS

STARTING THE DRILL. After laying out and previous to drilling, greatly enlarge the center holes with a center-punch to assist the starting of the drill. Start the hole with drill point in the enlarged center, using hand feed until a reasonable dimple is made in the work. Observe if this is central with the scribed circle, and if not central use center gouge, as in Fig. 14, and repeat until accurate.

TO DRAW A DRILL. When starting a drill it often has a tendency to slide or crowd off to one side. Where it is essential that the drilled hole coincide or center with some previously scribed circle or layout, the drill

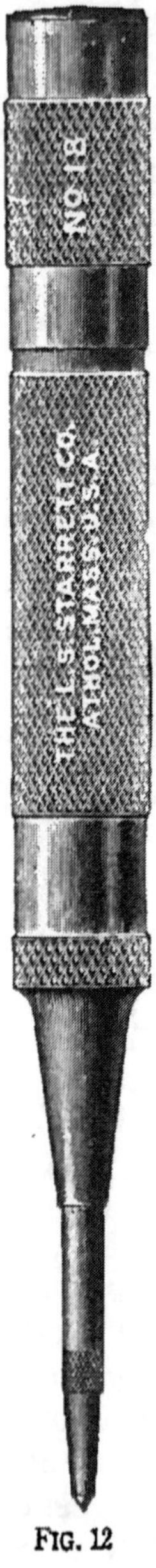

FIG. 12

must be brought back into the correct position. This is accomplished by the use of a small gouge-pointed chisel, sometimes called a center chisel, and the process is termed, "drawing the drill." First, note toward which side of the small dimple left by the drill-point it is necessary to shift the drill. Then chisel a small groove in that side of the dimple.

If the start is very eccentric, several chisel grooves may be necessary; whereas, if only slightly eccentric, a mere touch of the chisel will often suffice. It is readily seen that the drill is made to cut more easily where the grooves are, and therefore the natural resistance of the opposite side pushes the drill toward the side cut by the gouge-pointed chisel. Drill drawing can only be done previous to reaching the full diameter of cut.

HOLDING THE WORK. Carelessness in holding the work is responsible for many drilling accidents. If no special holding device is available, the work should be held in a drilling vise, clamped directly to the drilling-machine table, or clamped to an angle iron. Fig. 15 illustrates a method of holding the work safely. When once the work is clamped in position on the drilling-machine table, adjust the table to center the located hole with the drill rather than reclamp the work.

HOLDING THE DRILL. In Fig. 16, at A, the drill is shown held di-

FIG. 13

rectly in the spindle. This is a good method if several holes of the same diameter are to be drilled at a single setting. When frequent changing of the drill is necessary, as in drilling holes of numerous sizes, using a single-spindle machine, some form of quick-acting collett chuck should be used. The changes can then be made *without stopping the machine.*

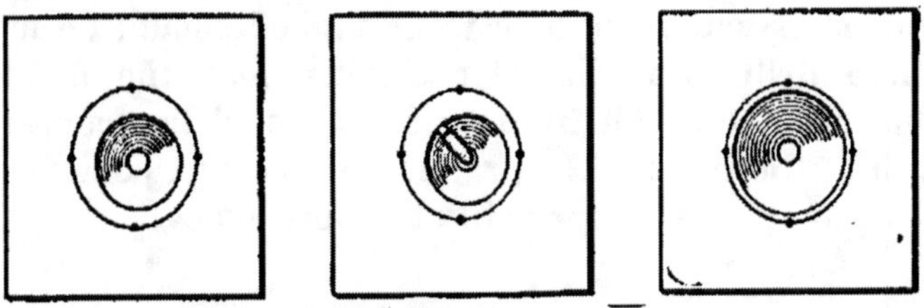

FIG. 14

DRILLING FOR REAMER. When it is essential that the holes be of an exact standard diameter, it is customary to use a drill somewhat smaller than the given diameter, and afterward ream the holes to standard size. The amount left for reaming depends upon whether one or two reaming operations are necessary, and whether or not the reaming is to be done directly in the drilling machine. If the drilling is done through jig bushings and the holes are short as compared to their diameter,

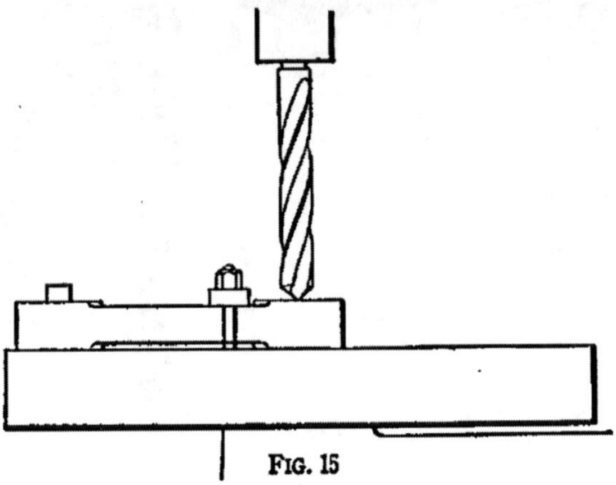

FIG. 15

a single reaming operation will often suffice. If the holes are relatively long, the drill should be 1/64" to 1/32" smaller than the finished hole diameter, to allow for passing a machine reamer 0.005" small through the hole which is afterward hand-reamed. This method gives results as accurate as any, except by grinding, and is accepted practice for good work.

DRILLING FOR TAPPING. Where a full thread depth is essential the hole to be tapped should be made with a drill of a diameter smaller than the nominal diameter of the bolt by an amount equal to double the depth of the thread. In practice the nearest commercial size of drill is listed for drilling tapped holes.

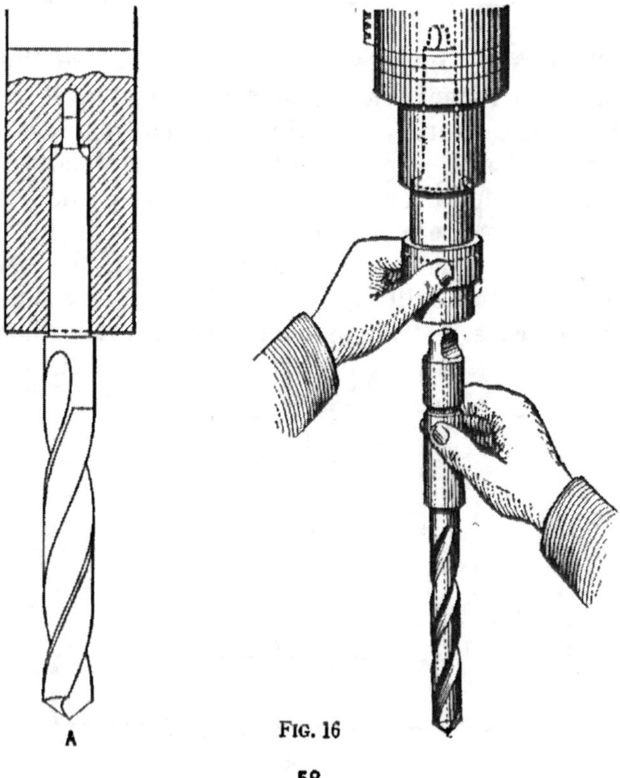

A

FIG. 16

Letter Sizes of Drills — Table 4

Diameter Inches	Decimals of 1 Inch	Diameter Inches	Decimals of 1 Inch
A 15/64	.234	N	.302
B	.238	O 5/16	.316
C	.242	P 21/64	.323
D	.246	Q	.332
E 1/4	.250	R 11/32	.339
F	.257	S	.348
G	.261	T 23/64	.358
H 17/64	.266	U	.368
I	.272	V 3/8	.377
J	.277	W 25/64	.386
K 9/32	.281	X	.397
L	.290	Y 13/32	.404
M 19/64	.295	Z	.413

Sizes of Tap Drills — Table 5

Tap Diameter	Threads per Inch	Drill for V Thread	Drill for U. S. Standards	Drill for Whitworth
1/4	16, 18, 20	5/32 5/32 11/64	3/16	3/16
9/32	16, 18, 20	3/16 13/64 13/64		
5/16	16, 18	7/32 15/64	1/4	15/64
11/32	16, 18	1/4 17/64		
3/8	14, 16, 18	1/4 9/32 9/32	9/32	9/32
3/32	14, 16, 18	19/64 21/64 21/64		
7/16	14, 16	21/64 11/32	11/32	11/32
15/32	14, 16	23/64 3/8		
1/2	12, 13, 14	3/8 25/64 25/64	13/32	3/8
9/16	12, 14	7/16 29/64	7/16	
5/8	10, 11, 12	15/32 1/2 1/2	1/2	1/2
11/16	11, 12	9/16 9/16		
3/4	10, 11, 12	19/32 5/8 5/8	5/8	5/8
13/16	10	21/32		
7/8	9, 10	45/64 23/32	23/32	23/32
15/16	9	49/64		
1	8	13/16	27/32	27/32

See also pages 78, 176 and 177.

Handy Equivalent Tables
Made of Spring Steel

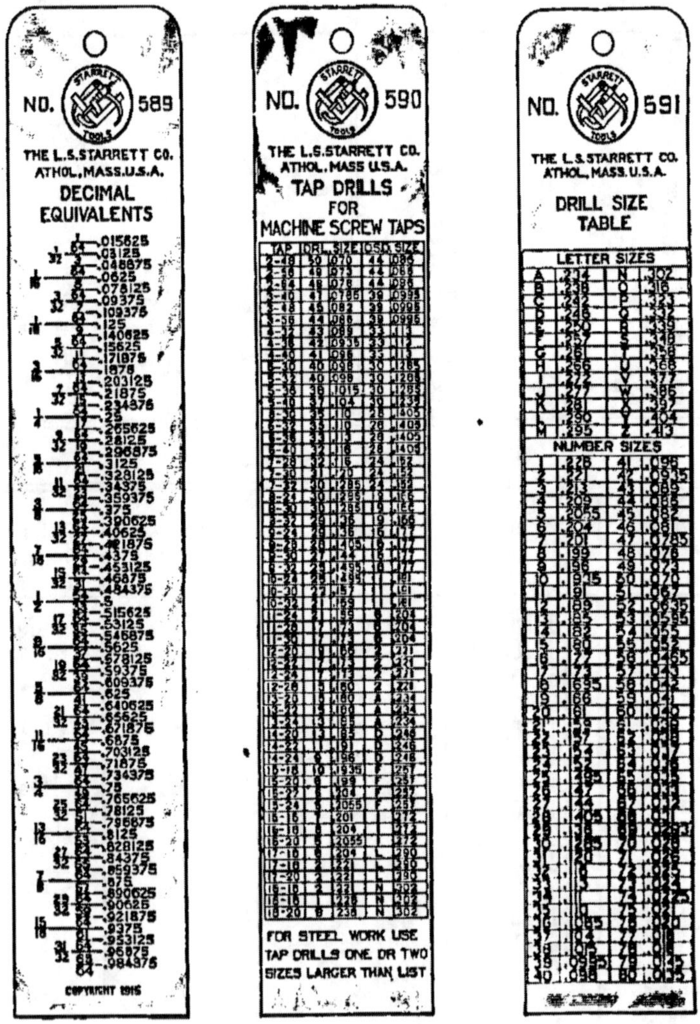

SIZES OF TAP DRILLS. Because of the large number of screw thread standards in use, many tables would be required to cover all selections of tap drills.

The sizes of tap drill for all pitches of V threads may be found by the following formula.

$$\text{Tap drill} = D - \frac{1.400}{T}$$

in which T = number of threads per inch
D = dia. of tap or thread

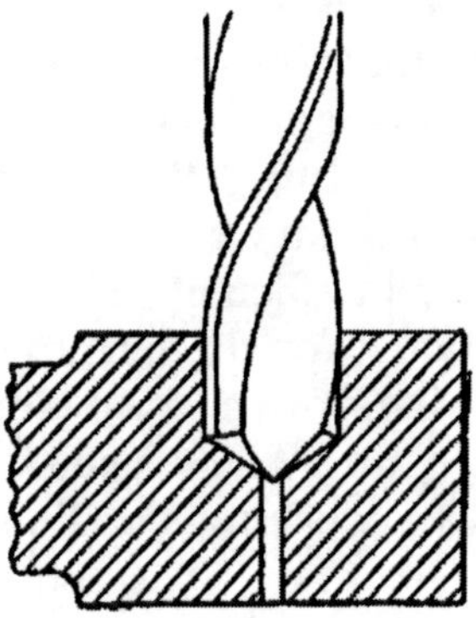

FIG. 17

EXAMPLE.—What diameter of tap drill should be used for a ¾ × 10 tap?

$$\text{Tap drill} = .75 - \frac{1.400}{10}$$
$$= .75 - .14$$
$$= .61$$

NOTE. For U. S. Standard threads use same formula, but 1.3 should be used in place of 1.4.

DRILLING LARGE HOLES. Twist drills are sold, ranging in size from No. 80 wire gage to four inches in diameter. As the drill increases in diameter the web is correspondingly thickened, and as the cutting edges at the web do not cut as effectively as they do outside the web thickness, considerable pressure is required to force the larger drills into the work at an efficient cutting feed. For this reason many workmen first drill a lead hole, using a drill whose diameter approximates the web thickness of the larger drill, as shown in Fig. 17. A lead hole will also assist in centering the drill upon an inclined surface. However, if the inclination is considerable it is necessary to butt mill or hand chip a spot giving sufficient surface to work upon. The practice of some firms is to use in place of a single large drill a relatively smaller one, afterward enlarging the hole by some method of counterboring at a much less expense for tools and at as rapid a production rate as by entire drilling.

BOLT HOLES. When the bolts are for holding purposes only and are not used for aligning the several pieces, it is customary to drill the holes through which the bolts pass somewhat larger than the bolt diameters. This allows for a variation in the bolt sizes and for inaccuracy in locating the centers.

DEEP HOLE DRILLING. Under this name may be classed the drilling of holes through the axes of spindles — lathe, milling-machine, and grinder — and that special line of drilling known as gun-barrel drilling. While for spindle drilling it is possible to use ordinary twist drills with extended shanks, it is customary in efficient drilling of this sort to use special drills designed for the purpose. Fig. 18 shows a special hollow drill often used for drilling axial holes in lathe spindles, and Fig. 19 shows the machine with the drill guides in working position.

Fig. 18

In all cases of deep-hole drilling it is better to rotate the work rather than the drill. The drill must be started exactly concentric with the axis of the machine. For this reason a starting-hole the exact diameter of the drill is first counterbored.

COUNTERBORING. There are many cases in which it is desirable to enlarge a hole throughout a portion of

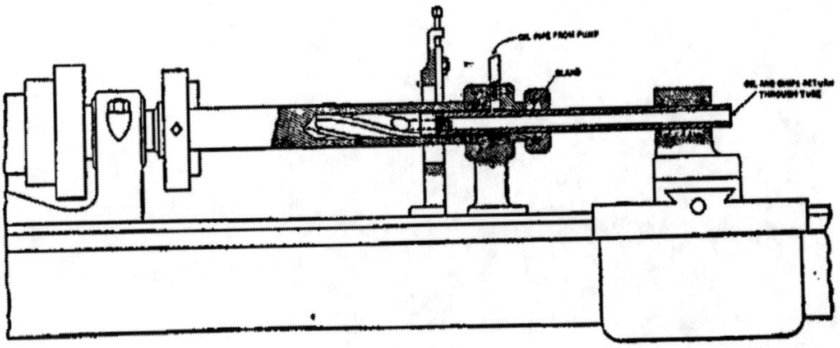

FIG. 19

its length. If a drill is used for this purpose there is no certainty that the two diameters will be concentric. The practice is to enlarge the already drilled hole by using a cutting tool having a pilot or leader to guide the cutting edges. This tool is known as a counterbore, and its use is termed counterboring. In Fig. 20 are shown the tool in operation and its purpose.

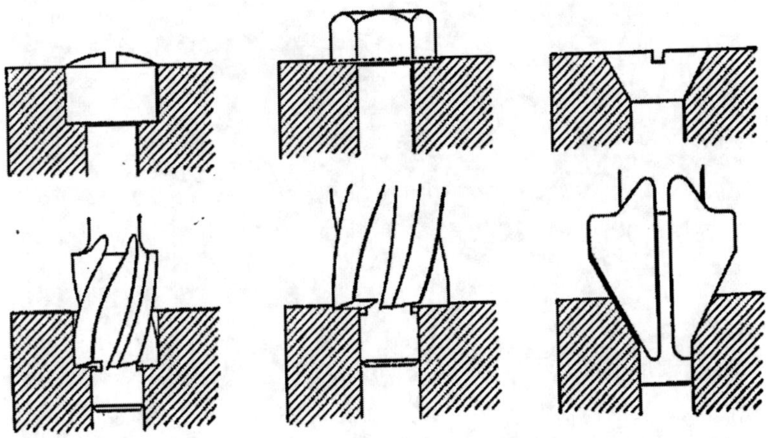

FIG. 20

THE LATHE

CARE OF THE LATHE. The engine lathe is capable of producing the largest variety of product of any of the machine-tool family. Especial attention should be given to applying a suitable machine oil to all the bearings, for improper lubrication of the wearing surfaces is one of the immediate causes of excessive wear. A medium-size flexible-bottom squirt can is best for this purpose, and oiling should be frequent on those bearings which are given the severest service, either from excessive pressure or from high-speed rubbing. All oil holes should be kept free and clean, and where possible should be protected from entering dirt. Those bearings, as, for example, the ways upon which the carriage moves, which by construction are hard to protect from dirt, should be frequently cleaned and reoiled. At least once a week the lathe should receive an all-over cleaning, and the bearings should be washed out with kerosene. A plugged oil hole prevents the proper lubrication of the bearing.

INDICATING AND ADJUSTING. Upon the condition of the centers, rests to a large degree the accuracy of the work produced. After attention to lubrication the competent workman proceeds to prepare and test the centers. Remove both centers and after cleaning them and the tapered holes note whether they return to their places with a successful fit. The "dead" or foot-stock center should have a hardened point to resist wear. The cone-points of the centers should be smooth and an exact sixty degrees. The centers should align with each other in the vertical and horizontal planes, and the "live" or head-stock cone-point should rotate truly concentric with its axis.

The trial and error method of adjusting the centers in alignment is to first bring the cone-points nearly into contact, and by adjusting the foot-stock frame upon its cricket bring them into as exact truth as is reasonably possible. With the foot-stock clamped in position to receive the work, surface the diameter of a trial piece for a length sufficient to allow testing its diameter at several places. If the diameter increases or decreases as the tool passes along the length of the work, readjust the foot-stock and repeat the test until the required

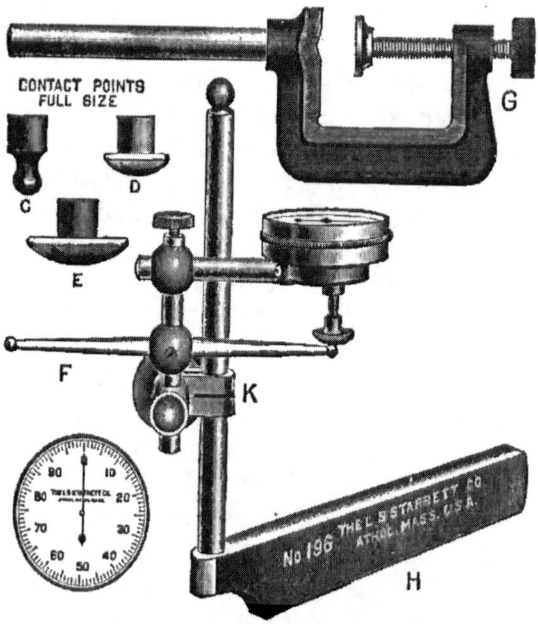

UNIVERSAL DIAL TEST INDICATOR
FIG. 21

degree of accuracy is obtained. To test the live center for concentricity, place in the tool-post a universal test-indicator, as shown in Fig. 21, with the feeler in touch with the cone-point. Rotate the head-stock spindle slowly by hand and note the dial. If the dial shows an eccentricity in excess of the allowed limits for the job to be done, the cone-point should be machined true. In cases where it is customary to have the live as well as the dead center hardened, the cone-point must be trued by some grinding attachment, as, for example, a tool-post grinding fixture. By many workmen the live center is left unhardened, and can be trued with a square nose-cutting tool, and afterward lightly filed to a smooth sur-

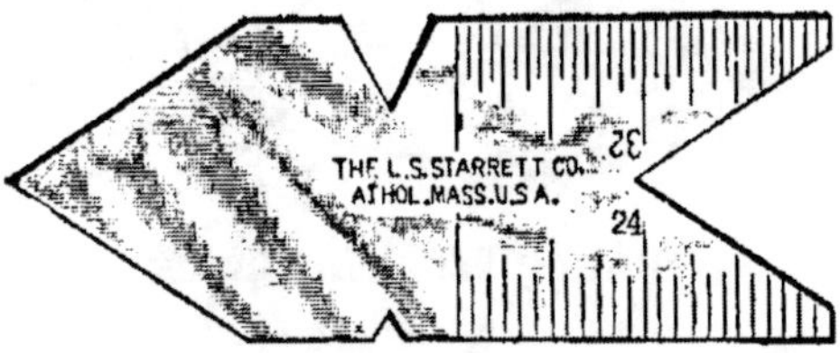

FIG. 22

face. To test either center for its proper cone-point angle use is made of a center gage, shown in Fig. 22.

TEST INDICATOR. This is a tool for indicating minute contact variations upon a graduated dial or upon

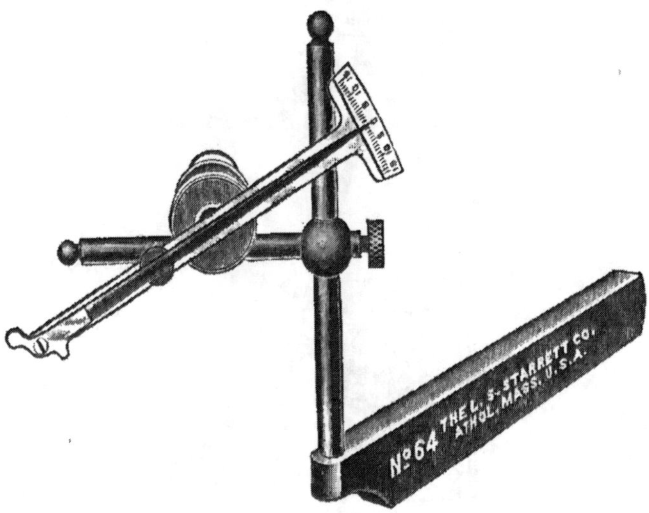

Truing Work in Chuck

Truing Jig on Face Plate

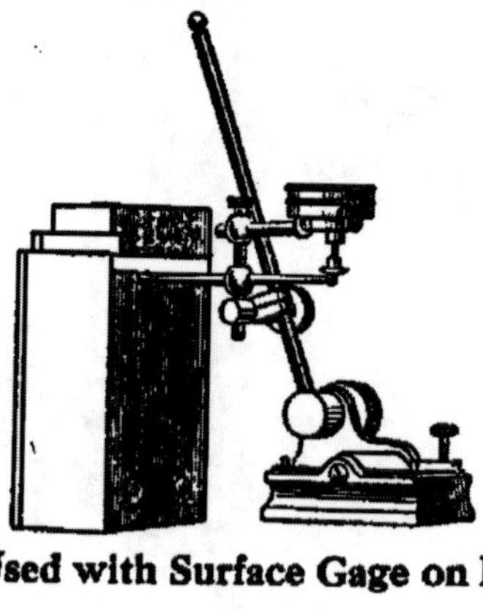

Indicator Used with Surface Gage on Bench Plate

a graduated arc. The graduations are usually one hundred in a complete circle with an easily read width of spacing. The instrument is built in such a way that one of these spaces represents a movement of the contact-point of 1/1000 inch.

Various mechanisms are employed for multiplying the movement of the contact-point, all of which are based upon a combination of short and long arm levers.

USE. The test-indicator may be used with advantage in any of the common machine tools, to indicate eccentricity in the lathe, milling machine, or grinding machine; to indicate uniformity of height in the planer, shaper, boring machine, or milling machine; to indicate parallelism, and to test for alignment in any machine.

WORK CENTERS. Most turned work is done upon the lathe centers, and it becomes necessary to provide suitable cavities in the work, coned to fit the cone-points. This is termed "centering the work," and consists in first locating the position of the cavities and afterward drilling and reaming them to form and size. Best practice in this respect is to use a combination drill and center reamer, as it insures exact concentricity in the drilled and reamed hole.

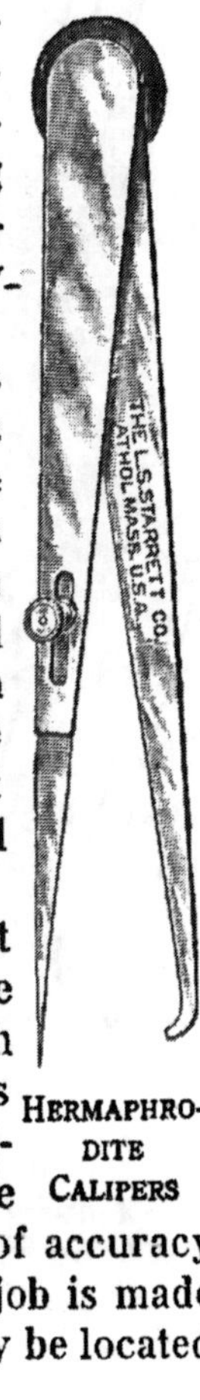

HERMAPHRODITE CALIPERS

LOCATING THE CENTERS. It is evident that the centers should be so located that the entire diameter of the turned job shall finish to size. Beside this, efficient turning demands that the chip taken shall be of practically uniform depth as the work rotates against the cutting tool. For these reasons some degree of accuracy in centering is necessary. Where the turned job is made from ordinary black bar stock, the centers may be located

LATHE TOOLS

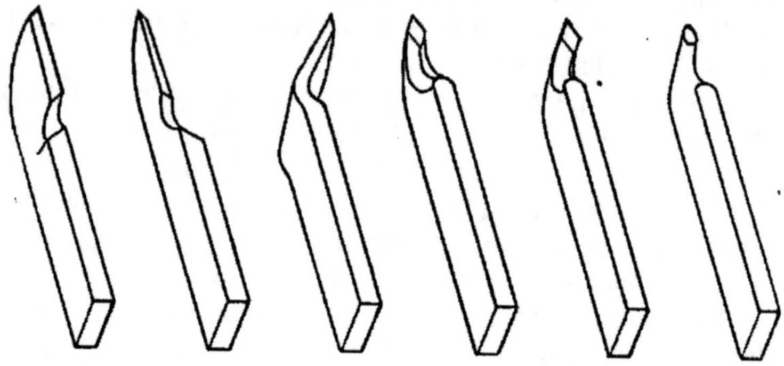

1 2 3 4 5 6

1 Left-hand Side Tool
2 Right-hand Side Tool
3 Right-hand Bent Tool
4 Right-hand Diamond Point
5 Left-hand Diamond Point
6 Round-nose Tool

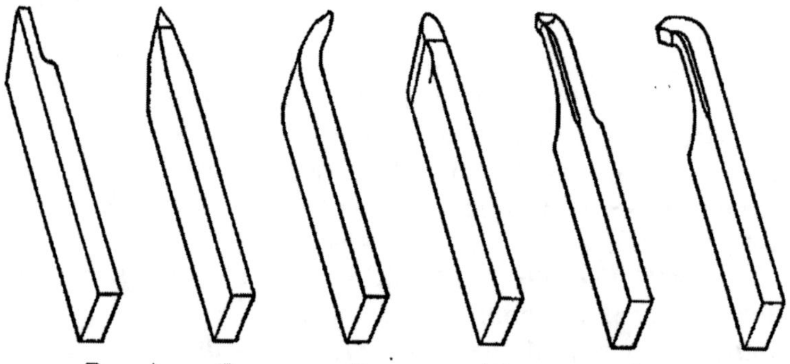

7 8 9 10 11 12

7 Cutting-off Tool
8 Threading Tool
9 Bent Threading Tool
10 Roughing Tool
11 Boring Tool
12 Inside Threading Tool

by scribing lines at an angle across the ends, using a combination square with a center head and the provided scriber. In place of this tool a hermaphrodite caliper may be used to scribe the ends of the stock. The center is located with a center-punch at the intersection of the scribed lines and the concentricity tested by spinning the bar upon the lathe centers. If necessary, the center-punch marks are shifted. If the piece is bent it must, after centering, be straightened to reasonable truth. For exact turned work the centers should afterward be lightly rereamed to correct the errors in their alignment due to the straightening of the bar.

When the job is to be turned from a forging, it is usual to roll the forging on straight edges and scribe lines across the ends, using a surface or height gage. In such cases the forging is so located with reference to the straight edges as to give a fair average of the surface errors due to forging. It is also usual to leave a greater excess of stock for finishing purposes upon a forging than upon rolled bar stock. When the centers are well located the holes may be drilled under a drill-press or in a hand-lathe, as convenient. Where much bar stock must be centered a special self-locating centering machine is often used.

LATHE TOOLS. A set of tools for use in the engine lathe is shown in the chart on page 70. While in common shop language all these are known as cutting tools, technically speaking, many of them separate the stock in a manner that is analogous to crowding off the metal rather than by pure cutting action. Cutting in its proper sense is a splitting action, and a properly ground and properly set cutting tool is a wedge in that it splits off the excess stock. Among the common lathe tools, the side tool and the diamond-point tool are the best examples of wedge or splitting action.

The nose of a cutting tool has several sides, two of

which come together at some angle to form a cutting edge. The angle formed by these surfaces must be sufficient for strength, and to furnish enough metal to conduct away the heat generated by the cutting action. For turning ordinary soft steel and soft gray iron an angle of sixty degrees is good practice. For harder materials the angle may be increased. In the case of forged lathe tools, the working end of the tool is forged upon the end of a short piece of square or rectangular bar stock. The length and size of the shank of the forged tool depend upon the size of chip and the machine used.

RAKE. The angle which the upper side of the tool makes with the horizontal is termed the rake. If the

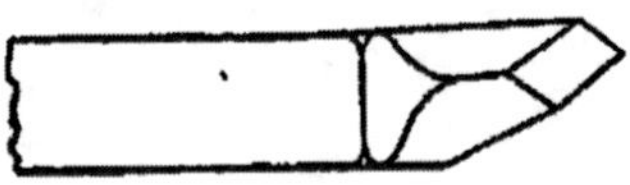

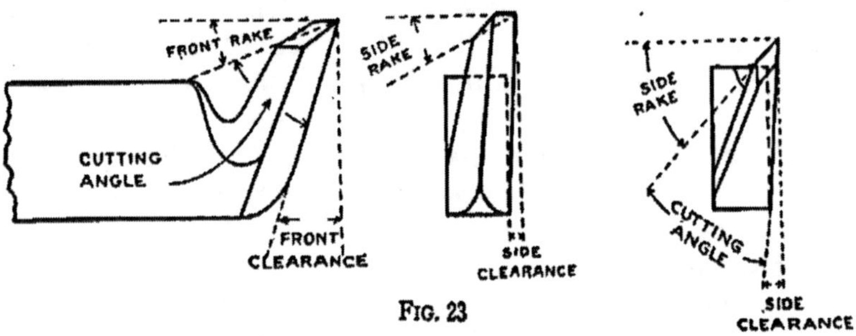

FIG. 23

slant is away from the work it is termed front rake; if in the direction of the axis of the work, it is termed side rake. A cutting tool may have its upper face forged and ground with either a front or a side rake or a combination of both. (See Fig. 23.)

CLEARANCE. By clearance is meant the angle which the under side of the tool makes with the vertical. As in the case of "rake" the clearance directly away from the axis of the work or lathe is termed front clearance, that along the axis of the work side clearance. With the tool in cutting position the clearances must be in any case not less than three degrees, and in most cases not more than ten degrees.

RIGHT-HAND TOOLS. These are tools having the rake, clearances, and cutting edges formed to turn or square from the right towards the left.

LEFT-HAND TOOLS. When the rake, clearances, and cutting edges are formed to cut from the left to the right the tool is known as a left-hand tool.

SETTING THE LATHE TOOL. It is very important that the lathe tool be properly set in relation to the axis of the work and the direction of the cut. While there are exceptions, notably that of the diamond point, lathe tools are usually set with the cutting point at the exact height of the axis of the lathe. In the case of the diamond point, the front clearance is usually forged to fifteen degrees or over. It is necessary, therefore, to set the point above the axis height to obtain a working clearance of not to exceed ten degrees. Unless the cutting tool has a bent shank it is usually set at right-angles to the surface of the work.

GRINDING LATHE TOOLS. Lathe tools made from carbon tool steel should be sharpened by grinding upon a *wet* emery-grinder, or upon an ordinary water-drip grindstone. If made from the newer high-speed steel the grinding should be upon a *dry* and rather coarse abrasive wheel. The grinder should have a suitable work-rest upon which to support the tool in sharpening the larger tools, or for resting the hands in the case of the smaller tools.

For purposes of safety, the work rest should be firmly and securely clamped as close as possible to the used face of the wheel. The grinding may be done upon the periphery of a disk-wheel or upon the sides of a cup-wheel, as desired. In any case the wheel should rotate to force the tool upon the rest rather than from it, and should run true and in balance. Efficient cutting depends very largely upon the correct sharpening, as well as the correct setting of the cutting tool, and great care should be taken when grinding a lathe tool to have the several faces true and making correct angles with each other. The manner of doing this is a pretty good index of the workman. The usual lathe-cutting tools have well-de-

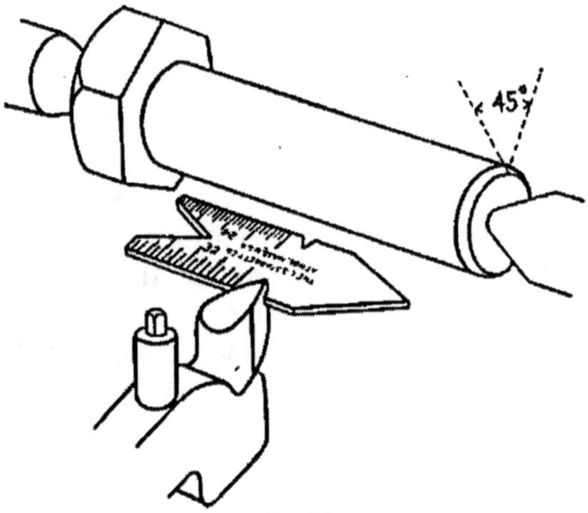

FIG. 24

fined cutting edges, and the angularity of the surfaces which meet to form the cutting edge can often be measured with a bevel protractor, and in the case of a sixty-degree angle the center gage is suitable. This tool is also used to test the angle when grinding a vee-pointed thread tool, as illustrated in Fig. 24.

TESTING THE CUTTING ANGLES. As the usual machine construction materials are not excessively hard, a cutting angle of not far from sixty degrees may be maintained on such tools as the side tool and the diamond point. In this case the angle can be tested by use of the usual center gage. Where cutting angles other than 60° are used, also for testing clearances, the universal Bevel Protractor is useful.

TOOL HOLDERS. The high cost of the materials used for modern cutting tools has resulted in the marketing of a variety of holders designed to hold cutting points. In this manner a large number of relatively inexpensive cutting points are made to interchange in a single shank or holder. One form of tool-holder is made to hold points forged in the regular forms shown in the chart, page 70. In some examples, however, the holders are made to carry short bits broken from square bar stock and afterward sharpened into some resemblance to the true forged shape. (See Fig. 25.)

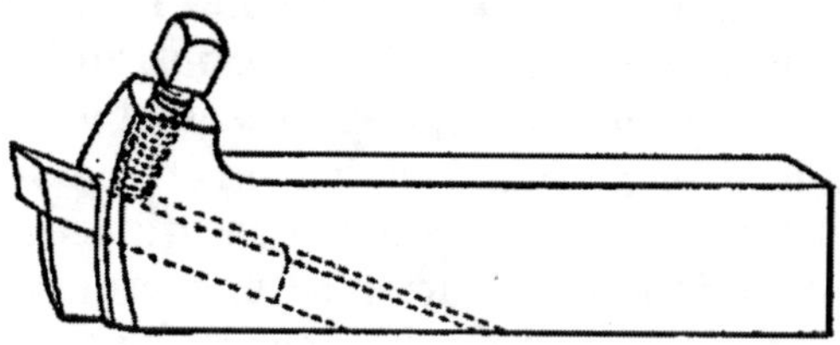

FIG. 25

MATERIALS FOR CUTTING TOOLS. These are known as carbon steel (tool steel), high-speed steel, and a new product of the electric furnace sold under the trade name of "Stellite." Carbon steel, or, as it was formerly termed, "tool steel," is high in carbon, eighty point to one hundred and twenty-five point, and when correctly heated and afterward plunged in cold water, hardens to a very high degree. Unfortunately for high-speed cutting the hardness is drawn at a comparatively low heat, and care must obtain not to overheat or blue it.

High-speed steel is a special steel having its composition alloyed with tungsten and perhaps vanadium or molybdenum. While heat treatment does not give it the exceeding hardness of tool or carbon steel, high-speed steel has the peculiar property of retaining its hardness at temperatures considerably in excess of those which readily soften tool steel. Tools made from high-speed steel are used at speeds, feeds, and cuts which heat the tools and chips to a dull red.

Stellite is a new cutting material composed of chromium, cobalt, and sometimes tungsten. It is cast into form and cannot be forged. Its hardness is equal to the diamond, and under favorable conditions marvelous turning may be done.

MANDRELS. Where the work is to be turned true with a hole through it, as, for example, turned pulleys, work-centers must be provided for holding it on the lathe centers. The common way is to force or drive into the work-hole a bar having center holes in its ends. This bar should be classed as a tool-room tool, and is properly known as a mandrel, although often called an arbor.

A standard set of mandrels varies in diameter and in length, according to the shop conditions. They are made of either tool steel hardened and ground true with the centers, or from soft machinery steel, case-carbonized and afterward ground. The ends for a short distance are reduced in diameter and provided with flats for clamping on the dog. Mandrels usually taper at the rate of 0.0005″ in an inch. The diameter of the hole fitted by the mandrel is stamped upon the larger end. As the quality of the work depends upon the truth of the mandrel it should be tested upon dead centers with a test-indicator before being used. To use, drive or force it into place, using a Mandrel press for forcing or a lead hammer for driving, carefully removing dirt, chips, or pieces of lead from the centers before placing the work in a lathe. Lathe drive with the usual lathe-dog as for any job done on the centers. Avoid forcing or driving the mandrel into a hole that is neither round nor straight. Also avoid scoring the mandrel with the cutting tool.

SCREW THREAD CUTTING. A screw thread is a helical groove cut or formed into the surface of a bar, rod, or bolt, or inside a nut. For ordinary machine screws, bolts, studs, etc., the threads are made with special tools called threading dies. These are screwed upon the bolt, screw, or stud to be threaded by rotating either the work or the die. Threading dies are used both by hand and in power-driven machines.

SCREW THREADS. There are numerous screw-thread standards in more or less general use. The so-called United States standard is in this country the more generally accepted one, and is therefore illustrated in Fig. 26 and Table 6. It will be noted that in addition to a definite form of thread cross-section each diameter has a specified number of threads per inch of length. The United States standard thread, when sectioned, shows a truncated sixty degrees triangle with the space and the land alike.

PITCH AND LEAD. Pitch in a thread is the distance measured from the center of one thread to the center of an adjacent thread. If the screw thread is a single helix, the lead is equal to pitch. If the helix is double, the lead is double the pitch. While strictly speaking pitch is the reciprocal of the number of threads per inch, as, for example, 1/7" pitch for a screw thread 7 per linear inch, shop men speak of it as 7 pitch, written, 7 P.

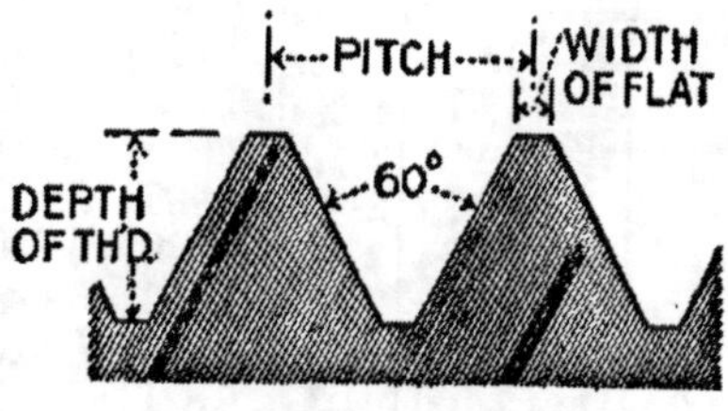

FIG. 26

THREADING IN A LATHE. When screw threads are cut in an engine lathe, the point of the cutting tool is shaped to the exact form of the spaces between threads. By means of a lead screw and a train of gearing the tool is compelled to move along the axis of the work at a

U. S. Standard Screw Threads — Table 6

Diameter	No. of Threads per Inch	Diameter at Root of Thread	Diameter of Tap Drill	Area in Sq. Inches: Of Bolt	Area in Sq. Inches: At Root of Thread	Dimensions of Nuts and Bolt Heads				
1/4	20	0.185	13/64	0.049	0.026	1/2	0.578	0.707	1/4	1/4
5/16	18	0.240	1/4	0.076	0.045	19/32	0.686	0.840	5/16	19/64
3/8	16	0.294	5/16	0.110	0.068	11/16	0.794	0.972	3/8	11/32
7/16	14	0.345	23/64	0.150	0.093	25/32	0.902	1.105	7/16	25/64
1/2	13	0.400	27/64	0.196	0.126	7/8	1.011	1.237	1/2	7/16
9/16	12	0.454	15/32	0.248	0.162	31/32	1.119	1.370	9/16	31/64
5/8	11	0.507	17/32	0.307	0.202	1 1/16	1.227	1.502	5/8	17/32
3/4	10	0.620	41/64	0.442	0.302	1 1/4	1.444	1.768	3/4	5/8
7/8	9	0.731	3/4	0.601	0.419	1 7/16	1.660	2.033	7/8	23/32
1	8	0.838	55/64	0.785	0.551	1 5/8	1.877	2.298	1	1 3/16
1 1/8	7	0.939	31/32	0.994	0.694	1 13/16	2.093	2.563	1 1/8	29/32
1 1/4	7	1.064	1 3/32	1.227	0.893	2	2.310	2.828	1 1/4	1
1 3/8	6	1.158	1 7/32	1.485	1.057	2 3/16	2.527	3.093	1 3/8	1 3/32
1 1/2	6	1.283	1 11/32	1.767	1.295	2 3/8	2.743	3.358	1 1/2	1 3/16
1 5/8	5 1/2	1.389	1 27/64	2.074	1.515	2 9/16	2.960	3.623	1 5/8	1 9/32
1 3/4	5	1.490	1 17/32	2.405	1.746	2 3/4	3.176	3.889	1 3/4	1 3/8
1 7/8	5	1.615	1 21/32	2.761	2.051	2 15/16	3.393	4.154	1 7/8	1 15/32
2	4 1/2	1.711	1 49/64	3.142	2.302	3 1/8	3.609	4.419	2	1 9/16
2 1/4	4 1/2	1.961	2 1/64	3.976	3.023	3 1/2	4.043	4.949	2 1/4	1 3/4
2 1/2	4	2.175	2 15/64	4.909	3.719	3 7/8	4.476	5.479	2 1/2	1 15/16
2 3/4	4	2.425	2 31/64	5.940	4.620	4 1/4	4.909	6.010	2 3/4	2 1/8
3	3 1/2	2.629	2 11/16	7.069	5.428	4 5/8	5.342	6.540	3	2 5/16
3 1/4	3 1/2	2.879	2 15/16	8.296	6.510	5	5.775	7.070	3 1/4	2 1/2
3 1/2	3 1/4	3.100	3 11/64	9.621	7.548	5 3/8	6.208	7.600	3 1/2	2 11/16
3 3/4	3	3.317	3 3/8	11.045	8.641	5 3/4	6.641	8.131	3 3/4	2 7/8
4	3	3.567	3 5/8	12.566	9.963	6 1/8	7.074	8.661	4	3 1/16
4 1/4	2 7/8	3.798	3 27/32	14.186	11.340	6 1/2	7.508	9.191	4 1/4	3 1/4
4 1/2	2 3/4	4.028	4 3/32	15.904	12.750	6 7/8	7.941	9.721	4 1/2	3 7/16
4 3/4	2 5/8	4.255	4 5/16	17.721	14.215	7 1/4	8.374	10.252	4 3/4	3 5/8
5	2 1/2	4.480	4 9/16	19.635	15.760	7 5/8	8.807	10.782	5	3 13/16
5 1/4	2 1/2	4.730	4 13/16	21.648	17.570	8	9.240	11.312	5 1/4	4
5 1/2	2 3/8	4.953	5 1/32	23.758	19.260	8 3/8	9.673	11.842	5 1/2	4 3/16
5 3/4	2 3/8	5.203	5 9/32	25.967	21.250	8 3/4	10.106	12.373	5 3/4	4 3/8
6	2 1/4	5.423	5 1/2	28.274	23.090	9 1/8	10.539	12.903	6	4 9/16

COURTESY OF "MACHINERY"

See also pages 55, 56, 168 and 169.

definite rate of advance as the work rotates. As the train of gears usually furnished with an engine lathe can be changed to give different rates of advance, it is in this manner possible to cut threads of a large variety of pitches. In practice a set of several gears having different numbers of teeth are furnished with each lathe. Those furnished will usually provide for cutting all the threads within the usual range of the lathe with which they come. These are known as "change gears," and their use is obvious.

SELECTING CHANGE GEARS. Given the number of threads per linear inch to be cut and the number of threads per linear inch of the lead screw, the problem is to select gears giving the desired ratio of cut to lead screw. For example, it is desired that single seven threads per linear inch shall be cut upon a 1¼-inch bolt, and it is found by scaling that the lathe lead screw has single five threads per linear inch. The ratio of cut to lead screw is then that of seven to five (7/5). The change gears selected should, therefore, be as seven is to five. If both members of a fraction are multiplied by the same number, the ratio is not changed. This allows of raising the fraction to suit the gears which are in the set furnished; for example, $\frac{7}{5} \times \frac{5}{5} = \frac{35}{25}$. If gears having thirty-five teeth and twenty-five teeth, respectively, are found in the furnished set, the selection of these gears will give, when rightly placed, the desired tool advance for cutting seven threads per linear inch.

The directions above refer to the most simple form of lathe. Various lathe manufacturers have introduced different arrangements of the gearing, but with any lathe the above procedure will give correct results if it is first determined what number of threads per inch will be cut if gears of the same number of teeth are placed on spindle stud and lead screw. This number called the

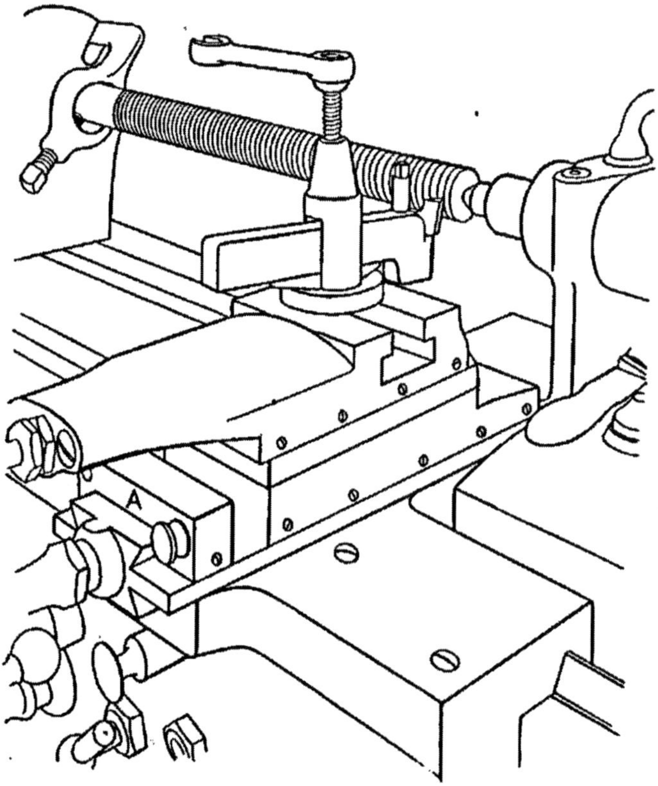

Lathe Set Up for Thread Cutting
Note Thread Stop at A

"lathe screw constant" should then be considered as being the number of teeth on the lead screw gear even though it is not the actual number.

PLACING THE CHANGE GEARS. The common engine lathe has projecting through its headstock a shaft known as the "stud." This projects a sufficient distance

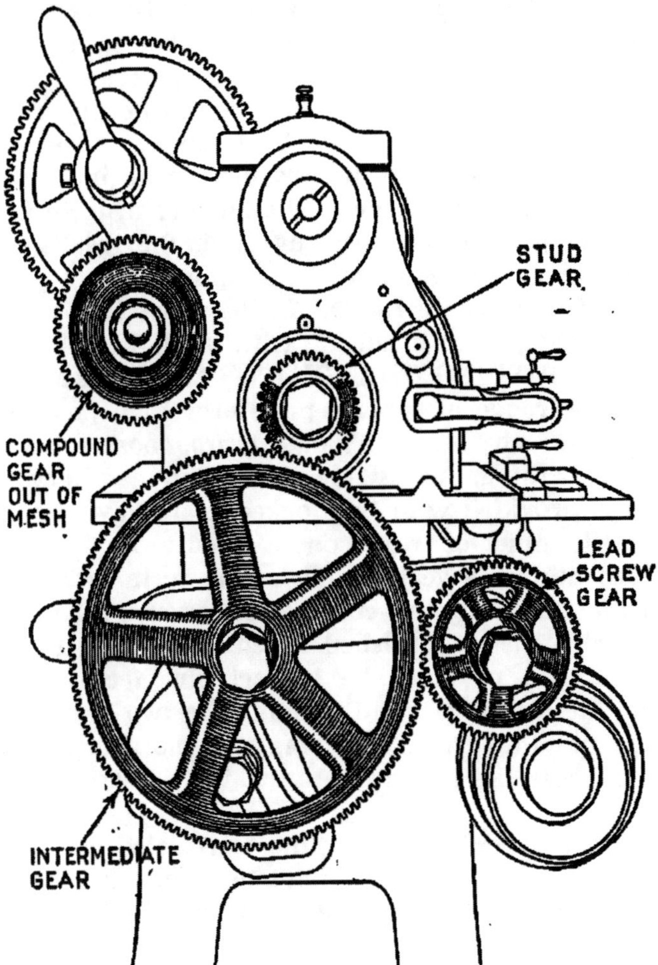

SIMPLE TRAIN OF GEARS FOR THREAD CUTTING

to allow of mounting gearing and usually the upper cone for the feed belt. Gears mounted or to be mounted upon this projecting stud are termed "stud gears." Those mounted upon the projecting end of the lead screw are known as lead gears. When the number of threads to be cut is more per linear inch than that of the lead screw, the smaller of the selected gears is placed upon the "STUD" and the larger upon the lead screw. In the example, the 25-tooth gear would be placed on the stud and the 35-tooth gear on the lead screw. Reverse the order if the number of threads per linear inch is less than that of the lead screw. The number of teeth in the large idler gear has no bearing upon the results, as it simply conveys the motion of the upper or stud gear to the lower or lead-screw gear. In the above it is assumed that the stud rotates in unison with the lathe spindle.

COMPOUNDING THE GEARS. As a means of enlarging the range of threads per linear inch possible to be cut with any set of change gears, most lathes are provided with an adjustable compound auxiliary stud which is provided with two locked gears having a ratio each to the other of two to one. As an example of their use, assume that a gear having ninety teeth was needed upon the lead screw to cut a given number of threads. If the set of gears furnished failed to provide a ninety gear, but did provide one of forty-five teeth, placing this on the lead screw and meshing the two to one compound stud into the train completes the desired ratio, and advances the tool as if the 90-tooth gear had been used.

THREAD TOOL. Among the tools listed on page 70 is shown the ordinary threading-tool point. It is obvious that this or any other form of point must be formed and tested to give the correct form of thread. The point shown has sides at an angle with each other of sixty degrees. The point can therefore be tested with a center

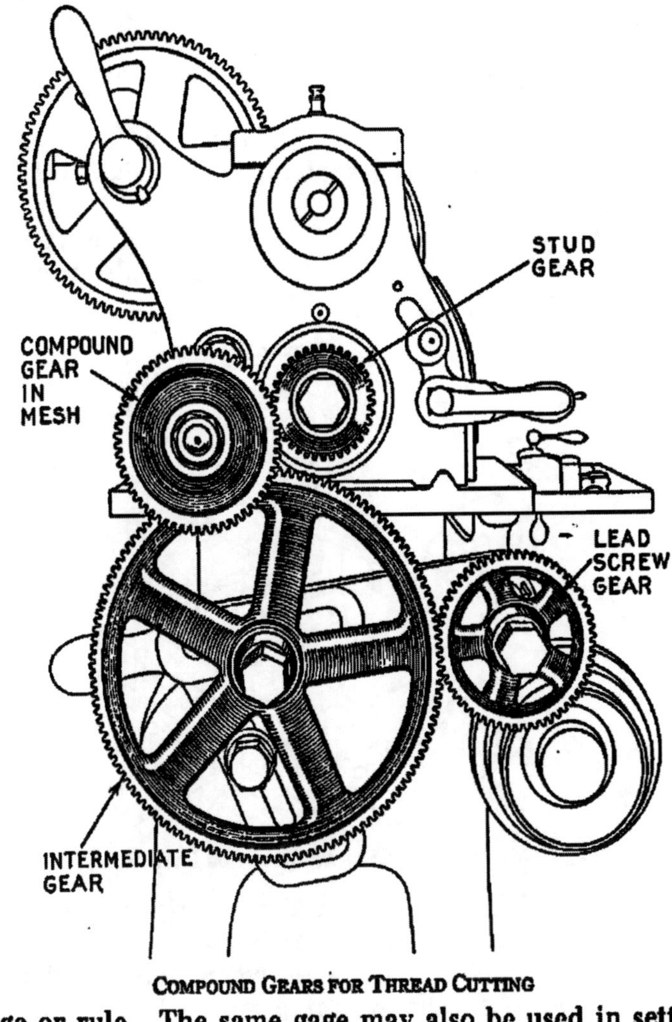

COMPOUND GEARS FOR THREAD CUTTING

gage or rule. The same gage may also be used in setting the tool square with the axis of the work (see page 74).

GRINDING THREAD TOOLS. It is important that the point of the thread tool shall conform to the outline of the groove between the adjacent threads, and that the surfaces below the cutting edge properly clear the stock being cut. When grinding a thread tool, particular care should be given to have the clearances sufficient for the lead of the thread.

SETTING THE TOOL. Set the tool point at the exact height of the lathe centers, and at right-angles to the axis of the lathe.

USES OF CUTTING LUBRICANT. Use lard oil when threading steel, wrought, and malleable iron. Cut the cast metals dry.

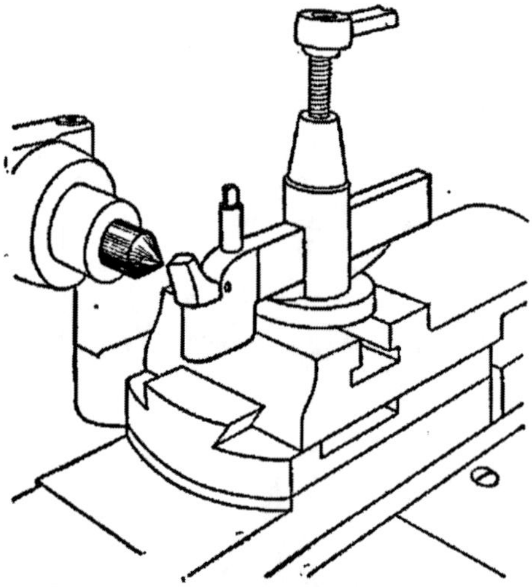

THREAD CUTTING TOOL SET AT HEIGHT OF LATHE CENTER

RIGHT AND LEFT THREADS. A right-hand thread results when the threading tool is advanced from right to left as it cuts. If the tool when cutting advances from left to right the resulting screw has a left-hand thread.

MEASURING AND TESTING SCREW THREADS. For ordinary purposes screw threads when cut are fitted to some threaded hole. This may be a hardened and ground gage, or may be an ordinary threaded nut, depending upon the accuracy of the work. Where the quality of the work demands special accuracy, or where standard threaded gages are not available, the thread is tested by measurements made with calipers. If the point of the thread tool has been carefully and exactly formed and accurately set in place, measuring the diameter at the root of the thread may give sufficiently accu-

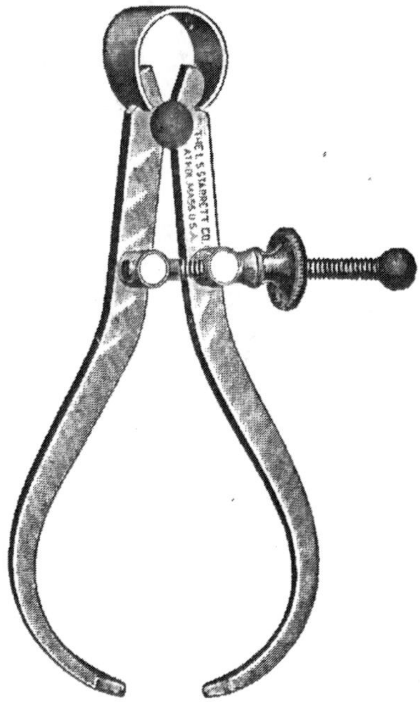

CALIPERS FOR TESTING THREADS

rate results, and this may be done with a set of thin point spring calipers. When greater accuracy than this is required, micrometers having special thread-measuring points are resorted to (see Fig. 27). In all this it is assumed that the thread tool is ground, set, and operated to give an exact thread outline.

MEASURING LATHE WORK. Work done in the engine lathe is of such a variety that a considerable list of measuring tools may be needed to cover all cases. Ordinarily, however, the diameter measurements can be made with spring calipers, micrometers, or some of the usual bar calipers. Cylindrical plug and ring gages, as well as limit snap gages, are also used for diameter measurements, and many of these may be used in measuring the shorter lengths. For the longer measurements of length, steel rules are provided with or without sliders. The more accurate measurements are usually made by using a micrometer.

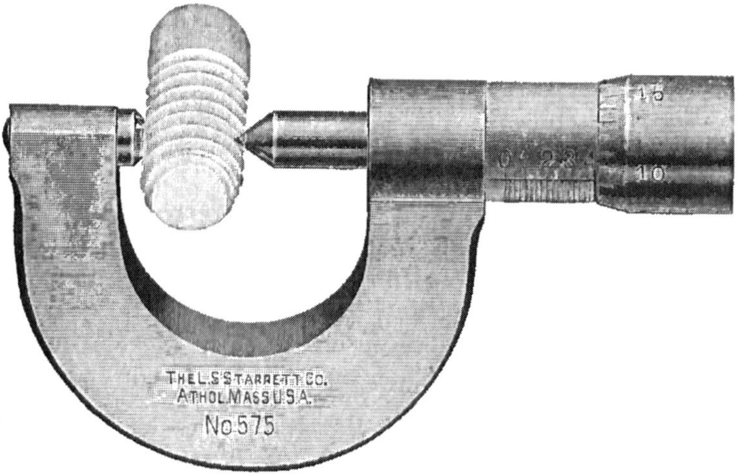

FIG. 27

TAPER TURNING. Where two parts are to fit firmly together when in use, as, for example, centers into lathe spindles, and it it desirable to have them easily removable, what are known as taper-fits are used. For this purpose several rates of change in diameter have become standards. Pages 87 and 88 give the more common standards. The Brown & Sharpe Standard is in general use for the spindle tapers in milling machines. The Morse taper is the one commonly used for all drills and drilling machinery. Either of these may be used for the tapered hole in lathe spindles, while some lathe manufacturers have established standards of their own.

Brown & Sharpe Taper Shanks—Table 7

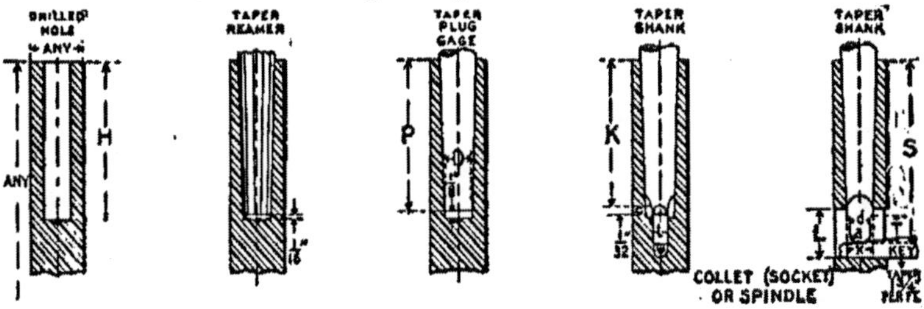

Taper per ft. is ½ in., except for No. 10 shank, where the taper is 0.5161 in. per ft.

Number of Taper	Diameter at End of Socket	Whole Length of Shank	Shank Depth	Diameter of Plug at Small End	Standard Plug Depth	Depth of Hole	End of Socket to Keyway	Length of Keyway	Width of Keyway	Length of Tongue	Thickness of Tongue
	A	*B*	*C*	*D*	*E*	*F*	*G*	*H*	-*K*	*L*	*M*
1	0.239	1 9/32	1 3/16	0.200	15/16	1 1/16	15/16	3/8	0.135	3/16	1/8
2	0.299	1 19/32	1 1/2	0.250	1 3/16	1 5/16	1 11/64	1/2	0.166	1/4	5/32
3	0.375	1 31/32	1 7/8	0.312	1 1/2	1 5/8	1 15/32	5/8	0.197	5/16	3/16
3	0.385	2 7/32	2 1/8	0.312	1 3/4	1 7/8	1 23/32	5/8	0.197	5/16	3/16
3	0.395	2 15/32	2 3/8	0.312	2	2 1/8	1 31/32	5/8	0.197	5/16	3/16
4	0.402	1 3/4	1 21/32	0.350	1 1/4	1 3/8	1 13/64	11/16	0.228	11/32	7/32
4	0.420	2 3/16	2 3/32	0.350	1 11/16	1 13/16	1 41/64	11/16	0.228	11/32	7/32
5	0.523	2 9/32	2 3/16	0.450	1 3/4	1 7/8	1 11/16	3/4	0.260	3/8	1/4
5	0.533	2 17/32	2 7/16	0.450	2	2 1/8	1 15/16	3/4	0.260	3/8	1/4
5	0.539	2 21/32	2 9/16	0.450	2 1/8	2 1/4	2 1/16	3/4	0.260	3/8	1/4
6	0.599	2 31/32	2 7/8	0.500	2 3/8	2 1/2	2 19/64	7/8	0.291	7/16	9/32
6	0.635	3 27/32	3 3/4	0.500	3 1/4	3 3/8	3 11/64	7/8	0.291	7/16	9/32
7	0.704	3 1/8	3 1/32	0.600	2 1/2	2 5/8	2 13/32	15/16	0.322	15/32	5/16
7	0.720	3 1/2	3 13/32	0.600	2 7/8	3	2 25/32	15/16	0.322	15/32	5/16
7	0.725	3 5/8	3 17/32	0.600	3	3 1/8	2 29/32	15/16	0.322	15/32	5/16
7	0.767	4 5/8	4 17/32	0.600	4	4 1/8	3 29/32	15/16	0 322	15/32	5/16
8	0.898	4 1/4	4 1/8	0.750	3 9/16	3 11/16	3 29/64	1	0.353	1/2	11/32
8	0.917	4 11/16	4 9/16	0.750	4	4 1/8	3 57/64	1	0.353	1/2	11/32
9	1.067	4 3/4	4 5/8	0.900	4	4 1/8	3 7/8	1 1/8	0.385	9/16	3/8
9	1.077	5	4 7/8	0 900	4 1/4	4 3/8	4 1/8	1 1/8	0.385	9/16	3/8
10	1.260	5 27/32	6 23/32	1.0446	5	5 1/8	4 27/32	1 5/16	0.447	21/32	7/16
10	1.289	6 17/32	6 13/32	1.0446	5 11/16	5 13/16	5 17/32	1 5/16	0.447	21/32	7/16
10	1.312	7 1/16	6 15/16	1.0446	6 7/32	6 11/32	6 1/16	1 5/16	0.447	21/32	7/16
11	1.498	6 25/32	6 21/32	1.250	5 15/16	6 1/16	5 25/32	1 5/16	0 447	21/32	7/16
11	1.531	7 19/32	7 15/32	1.250	6 3/4	6 7/8	6 19/32	1 5/16	0.447	21/32	7/16
12	1.797	8 1/16	7 15/16	1.500	7 1/8	7 1/4	6 15/16	1 1/2	0.510	3/4	1/2
13	2.073	8 11/16	8 9/16	1.750	7 3/4	7 7/8	7 9/16	1 1/2	0 510	3/4	1/2
14	2.344	9 9/32	9 5/32	2.000	8 1/4	8 3/8	8 1/32	1 11/16	0.572	27/32	9/16
15	2.615	9 25/32	9 21/32	2.250	8 3/4	8 7/8	8 17/32	1 11/16	0.572	27/32	9/16
16	2.885	10 3/8	10 1/4	2.500	9 1/4	9 3/8	9	1 7/8	0.635	15/16	5/8
17	3.156	....	....	2.750	9 3/4	9 7/8	....	....	....	....	...
18	3.427	....	....	3.000	10 1/4	10 3/8	....	....	....	....	...

Morse Standard Taper Shanks—Table 8

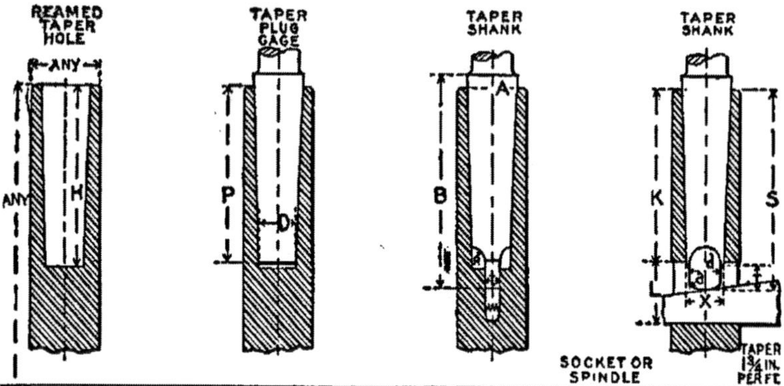

Number of Taper	Diameter of Plug at Small End	Dia. at End of Socket	Standard Plug Depth	Whole Length of Shank	Depth of Hole	End of Socket to Keyway	Length of Keyway	Length of Tongue	Thickness of Tongue	Width of Keyway	Shank Depth	Taper per Foot
	D	*A*	*P*	*B*	*H*	*K*	*L*	*T*	*t*	*W*	*S*	
0	0.252	0.356	2	2 11/32	2 1/32	1 15/16	9/16	1/4	5/32	0.160	2 7/32	.625
1	0.369	0.475	2 1/8	2 9/16	2 3/16	2 1/16	3/4	5/16	13/64	0.213	2 3/8	.600
2	0.572	0.700	2 9/16	3 1/16	2 5/8	2 1/2	7/8	3/8	1/4	0.260	2 7/8	.602
3	0.778	0.938	3 3/16	3 3/4	3 1/4	3 1/16	1 1/16	7/16	5/16	0.322	3 9/16	.602
4	1.020	1.231	4 1/16	4 3/4	4 1/8	3 7/8	1 1/4	1/2	15/32	0.478	4 1/2	.623
5	1.475	1.748	5 3/16	6	5 1/4	4 15/16	1 1/2	5/8	5/8	0.635	5 3/4	.630
6	2.116	2.494	7 1/4	8 5/16	7 3/8	7	1 3/4	7/8	3/4	0.760	8	.626
7	2.750	3.270	10	11 5/8	10 1/8	9 1/2	2 5/8	1 3/8	1 1/8	1.135	11 1/4	.625

Short Shanks

Number of Taper	D	A	P	B	H	K	L	T	t	W	S	Taper per Foot
0	0.271	0.356	1 5/8	1 31/32	1 21/32	1 17/32	5/8	1/4	3/16	0.195	1 27/32	.625
1	0.388	0.475	1 3/4	2 3/16	1 13/16	1 21/32	13/16	5/16	1/4	0.260	2	.600
2	0.600	0.700	2	2 9/16	2 1/16	1 27/32	1 3/16	7/16	3/8	0.387	2 3/8	.602
3	0.816	0.938	2 7/16	3 1/8	2 1/2	2 7/32	1 5/16	9/16	1/2	0.514	2 15/16	.602
4	1.062	1.231	3 1/4	4 1/16	3 5/16	2 31/32	1 1/2	5/8	5/8	0.639	3 13/16	.623
5	1.532	1.748	4 1/8	5 1/16	4 3/16	3 21/32	2	3/4	1	1.014	4 13/16	.630
6	2.201	2.494	5 5/8	7 1/16	5 3/4	5 1/16	2 3/4	1 1/8	1 1/4	1.266	6 3/4	.626
7	2.857	3.270	7 15/16	9 11/16	8 1/16	7 1/8	3 5/8	1 1/2	1 5/8	1.642	9 5/16	.625

The dimensions given above for regular (full length) Morse taper shanks are those which have been accepted as standard and are used by most manufacturers. In a recent catalogue of the Morse Twist Drill & Machine Co., however, a table is given in which the length of the tang and, consequently, the whole length of the shank is slightly increased. The increase in length, however, is so slight that it does not prevent the shank from fitting into the ordinary standard taper socket.

TURNING TAPERS. Ordinary tapers are rated at the amount which the diameter changes in a foot's length; as, for example, the Brown & Sharpe taper of ½ inch per foot. To turn a taper it is necessary to use a lathe provided with a taper attachment or to adjust the footstock of the engine lathe sufficiently off center to give

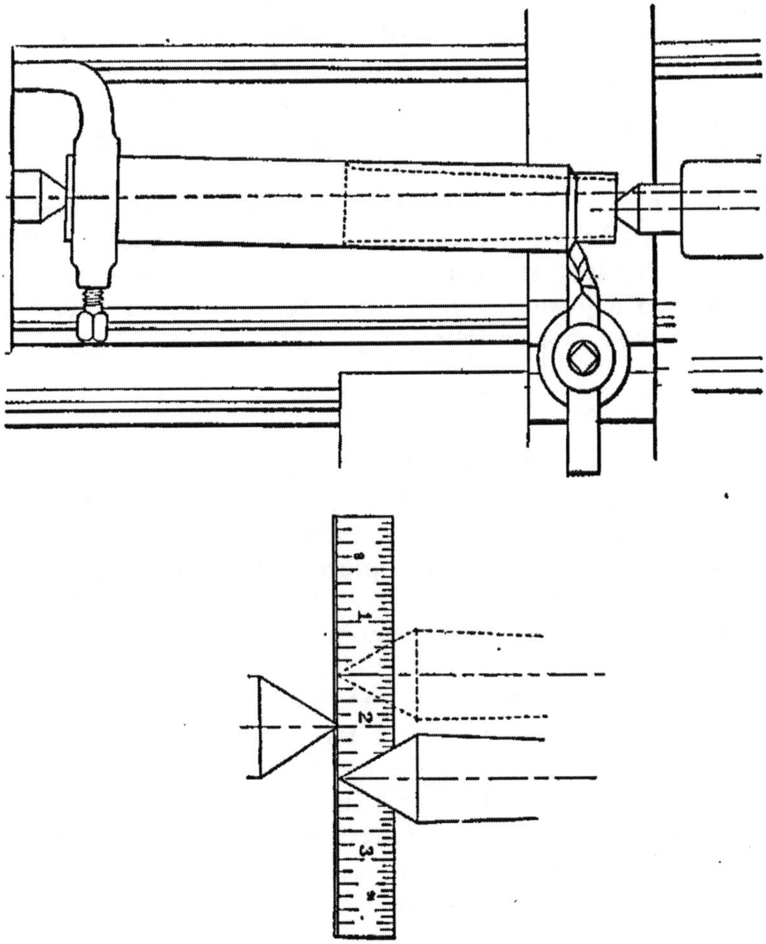

TAPER TURNING IN LATHE

the required rate of diameter change. As all taper attachments are graduated to read direct, they are easily set for the required taper. Adjustment of the foot-stock of an engine lathe is not so simple as the taper attachment. In setting the taper attachment, the axial distance the center points are apart is not important, while this distance must be considered in setting over the foot-stock of the lathe.

AMOUNT TO OFFSET CENTERS FOR GIVEN TAPER. If the distance the center points enter the work or the mandrel is ignored, the mandrel length can be considered as the distance apart of the center points. The calculation necessary to determine the distance which the centers shall be offset, is that of multiplying the length of the work or mandrel in feet by one-half of the required taper in inches. To turn a Brown & Sharpe taper on a piece of work nine inches long the problem would work out as follows:

$$\frac{.500}{2} \times \frac{9}{12} = 0.1875 = \frac{3}{16}$$

and the foot-stock would be set over $\frac{3}{16}$ inch.

In the above illustrative example both length and amount of taper are given, but the amount of taper is not always known. Suppose a piece is 8 inches long and a taper is to be turned on one end, the tapered portion to be 4 inches long. The difference in diameters of these 4 inches is to be ½ inch. How much must the tail stock be offset? If the taper is ½ inch in 4 inches it would be 1½ inches in a foot and the tail stock would be moved over one-half of 1½ inches or ¾ inch, if the piece were a foot long, but as it is only 8 inches or ⅔ of a foot long, the tail stock should be moved over ⅔ multiplied by ¾ or ½ inch. Had the piece been 18 inches long, the tail stock should be moved over $\frac{3}{2}$ multiplied by ¾ or 1⅛ inches.

It has been assumed for these simple calculations that the lathe centers merely touch the ends of the piece, thus making the length of the piece the same as the distance between centers. But in actual work the distance the centers enter the piece must be considered. The calculation should be as accurate as possible to avoid continually changing the tail stock to get a reasonably good taper fit. The necessity of considering the distance the center enters the piece depends somewhat upon its length. If the piece is very long, the actual taper will differ considerably from the calculated taper. If each center enters the piece one-fourth inch they would enter a total of one-half inch, and the length of the piece should be reduced by one-half inch in the calculation. While turning the taper the calipers should be used frequently so that it may be soon determined whether or not the tail stock is correctly placed.

For coning pulleys, set the foot-stock away from the operator when adjusting. In most taper work, however, the center is offset towards the operator.

SETTING THE TOOL. The tool-point should be set at the exact height of the axis of the lathe.

TESTING THE TURNED TAPER. To test the taper as it is turned, ground, or filed, it should be pressed lightly into a standard tapered hole and worked back and forth sufficiently to mark the places where bearing occurs. If the work has been lightly covered with some marking pigment, the bearing points will be more distinct. Care, however, must obtain that the coating is not sufficient to smooch, as it will deceive the workman. Adjust taper-setting until a correct fit is obtained.

ECCENTRIC TURNING. While for the most part the lathe is used for work exactly concentric with the axis, it can be used for turning work not concentric with the axis. Work of this sort is termed "eccentric," and an example of such work is seen in the eccentrics which

Amount of Taper in a Given Length, When the Taper per Foot is Known — Table 9

Length of Taper	Taper per Foot										
	1/16	3/32	1/8	1/4	3/8	1/2	0.600	5/8	3/4	1	1 1/4
1/32	.0002	.0002	.0003	.0007	.0010	.0013	.0016	.0016	.0020	0.0026	0.0033
1/16	.0003	.0005	.0007	.0013	.0020	.0026	.0031	.0033	.0039	0.0052	0.0065
1/8	.0007	.0010	.0013	.0026	.0039	.0052	.0062	.0065	.0078	0.0104	0.0130
3/16	.0010	.0015	.0020	.0039	.0059	.0078	.0094	.0098	.0117	0.0156	0.0195
1/4	.0013	.0020	.0026	.0052	.0078	.0104	.0125	.0130	.0156	0.0208	0.0260
5/16	.0016	.0024	.0033	.0065	.0098	.0130	.0156	.0163	.0195	0.0260	0.0326
3/8	.0020	.0029	.0039	.0078	.0117	.0156	.0187	.0195	.0234	0.0312	0.0391
7/16	.0023	.0034	.0046	.0091	.0137	.0182	.0219	.0228	.0273	0.0365	0.0456
1/2	.0026	.0039	.0052	.0104	.0156	.0208	.0250	.0260	.0312	0.0417	0.0521
9/16	.0029	.0044	.0059	.0117	.0176	.0234	.0281	.0293	.0352	0.0469	0.0586
5/8	.0033	.0049	.0065	.0130	.0195	.0260	.0312	.0326	.0391	0.0521	0.0651
11/16	.0036	.0054	.0072	.0143	.0215	.0286	.0344	.0358	.0430	0.0573	0.0716
3/4	.0039	.0059	.0078	.0156	.0234	.0312	.0375	.0391	.0469	0.0625	0.0781
13/16	.0042	.0063	.0085	.0169	.0254	.0339	.0406	.0423	.0508	0.0677	0.0846
7/8	.0046	.0068	.0091	.0182	.0273	.0365	.0437	.0456	.0547	0.0729	0.0911
15/16	.0049	.0073	.0098	.0195	.0293	.0391	.0469	.0488	.0586	0.0781	0.0977
1	.0052	.0078	.0104	.0208	.0312	.0417	.0500	.0521	.0625	0.0833	0.1042
2	.0104	.0156	.0208	.0417	.0625	.0833	.1000	.1042	.1250	0.1667	0.2083
3	.0156	.0234	.0312	.0625	.0937	.1250	.1500	.1562	.1875	0.2500	0.3125
4	.0208	.0312	.0417	.0833	.1250	.1667	.2000	.2083	.2500	0.3333	0.4167
5	.0260	.0391	.0521	.1042	.1562	.2083	.2500	.2604	.3125	0.4167	0.5208
6	.0312	.0469	.0625	.1250	.1875	.2500	.3000	.3125	.3750	0.5000	0.6250
7	.0365	.0547	.0729	.1458	.2187	.2917	.3500	.3646	.4375	0.5833	0.7292
8	.0417	.0625	.0833	.1667	.2500	.3333	.4000	.4167	.5000	0.6667	0.8333
9	.0469	.0703	.0937	.1875	.2812	.3750	.4500	.4687	.5625	0.7500	0.9375
10	.0521	.0781	.1042	.2083	.3125	.4167	.5000	.5208	.6250	0.8333	1.0417
11	0573	.0859	.1146	.2292	.3437	.4583	.5500	.5729	.6875	0.9167	1.1458
12	.0625	.0937	.1250	.2500	.3750	.5000	.6000	.6250	.7500	1.0000	1.2500

operate the valves of steam engines. If the work has a hole through it, as in the above example, the hole is first finished to required dimensions. A mandrel is then used for carrying the work on the centers. While the mandrel has been built on one set of centers exactly true with its axis, for eccentric turning it has a second set of centers which are offset the amount required for the eccentricity specified. In the case of eccentrics made solid with the

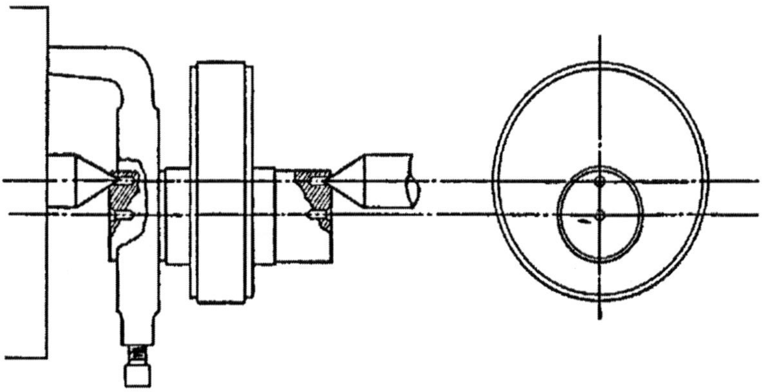

FIG. 28

shaft, the two sets of centers, one for turning the shaft and the other for finishing the eccentrics, are made side by side in the ends of the shaft, as shown in Fig. 28.

When the specified eccentricity is too extreme to allow both pairs of centers coming within the limits of the diameter of the shaft, special ends may be cast or forged on the ends of the work, and can afterward be machined off. In crank-shaft turning, special attachments are provided for the ends of the shaft. Special eccentric turning chucks may be made to hold the work.

CHUCKING. Chucking includes, not only the mounting of the work in the chuck, but performing the necessary operations on it while so held. The name "chuck" is given to a line of tools having a variety of form, all

of which are designed to hold work or tools upon the nose of a spindle. In general the heavier sorts are mounted upon a face-plate which screws upon the end of the spindle, while smaller sizes are fitted with a taper-shank which fits tightly into the tapered hole in the spindle. The smaller sizes are used for carrying tools, such as drills, also screws, studs, wire pins, etc.; and are known as drill-chucks.

The larger sizes are widely used for holding work for machine operations, and are sometimes called "work-chucks." On their face they are provided with adjusting jaws movable regularly to and from the center; these jaws are so designed that a considerable variety of work may be readily held and successfully worked upon with common cutting tools. The jaws are moved by means of screws or gears, and can be adjusted independently, the chuck being called an independent jaw-chuck; or, all the jaws may be made to move together, in which case it is known as a Universal chuck.

HOLDING THE WORK. The work must be clamped firmly in the chuck while being machined. Care must also be taken that the clamping of a slender piece is not so firm as to distort or spring it. If work slips, tools may be broken, and if held too tightly and sprung or crushed, the work is injured and in some cases entirely ruined.

TRUING THE WORK. Adjusting the chuck-jaws so that the work will run as true as desired is termed, "truing up the work." This is preliminary to any tooling which may be done on the job. Often this truing of the work can be accomplished by holding a piece of chalk to just touch the work, leaving a plain marking—this method is used when chucking rough pulleys for drilling out the hole in the hub. Where greater accuracy is required, the work is indicated with a Universal dial test indicator.

CHUCKING TOOLS. With the work located in the chuck it may be tooled with ordinary lathe tools, such as shown in the tool-chart (page 70), or it may be drilled with two, three, or four fluted twist drills, and reamed with machine reamers, or special shell bits and counterbores.

CHUCKS ON TURRET LATHES. In turret lathe-work, for bar-stock, the chuck is a part of the regular tool equipment; these chucks are often of special design, so made that they open and close by hand-operated levers or automatically-operated cams.

KNURLING. The surfaces of adjusting screws and small machine parts are often given a regular rough surface for easy gripping. In the machine shop this is done by using a tool known as a "knurl" or "knurling tool," which consists of one or more indented rollers or knurls mounted to rotate in some form of holder.

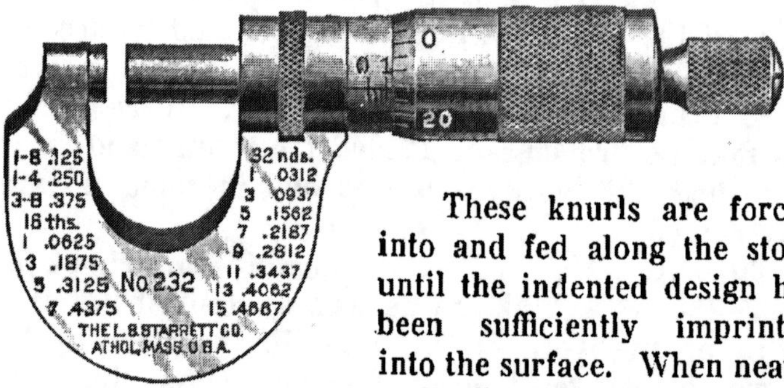

FIG. 29

These knurls are forced into and fed along the stock until the indented design has been sufficiently imprinted into the surface. When neatly and effectively done the results give a fine gripping surface and a rather pleasing effect to the eye. The knurling tool may be fed along the surface of the work by hand, but usually the power traverse feed is used. The process is repeated if one passage of the tool does not give sufficient depth.

Fig. 29 shows knurling on a micrometer.

TOOL-MAKING

Under the name "tools" are listed the various small or tool-room tools used either by hand or in various machines. So important has their use become that large industries are devoted to their manufacture, and most machine-building firms now buy their more common tools rather than maintain a tool-making plant of their own. For example, drills, reamers, milling cutters, counterbores, colletts, etc., are usually purchased in the open market. Every skilled machinist, however, should know the principles upon which such tools are made, and should be able to make any or all of them.

DRILLS. Drills are now largely of the twist type, and the most efficient are machined and milled from solid bar-stock, and for this purpose both carbon-tool steel and high-speed steel are being used. The prevailing type has a straight or a tapered holding shank, spiral-milled flutes and a cone-point with effective cutting lips as noted under drill sharpening. The flutes or lands taper slightly from full diameter size at the cone-point to several thousandths inch smaller at or near the holding shank. To prevent rubbing on the sides of a hole, the flutes are also cleared back from the front edge throughout their length. The grooves are milled with cutters having a form that gives the maximum chip capacity, yet leaves the cutting edge of the drill-lip a straight line.

Several makers of twist-drills increase the lead of the twist when milling the grooves; such drills are known as "increase twist" drills. The web is as thin as consistent with the required strength, and with some makers is thicker near the shank than at the point. Drills are carefully heat-treated, straightened, and ground to diameter.

REAMERS. The term "reaming" is given to the process of enlarging a drilled hole. Reamers are of two well-defined types, known as "fluted" reamers and "rose" reamers. The fluted reamer is one having numerous flutes on the circumference of the cutting portion of the tool. In other words, the cutting is done on the circumference instead of at the end, as with a drill.

The number of flutes on the surface of a reamer varies with the diameter, and with some makes the number of flutes is greater for a given diameter when the reamer is to be used in a machine instead of for hand reaming.

As its name implies, a fluted hand reamer is made for hand use, and is seldom called upon to enlarge a hole more than .007" for any diameter, and not more than .003" in the smaller sizes.

In the case of machine or lathe reamers, the length of the flutes for any given diameter is fifty per cent less than the standard length for hand reamers. The depth of flute is usually somewhat in excess of that of hand reamers. In most cases machine reamers are used for enlarging drilled holes to a diameter which only allows sufficient stock for hand reaming. When the holes are not to exceed a diameter in length, machine reamers may be used for finishing the drilled hole to its full diameter; but when straight, round, accurate holes are to be of exact diameter the better practice is to first drill 1/32" to 1/16" under size, enlarge to hand reaming size with a machine reamer, and then carefully hand ream to exact size.

ECCENTRIC FLUTES. Formerly fluted reamers had an odd number of flutes, such as nine or eleven. Although this method eliminated chattering to some extent, it had the disadvantage of making it difficult to caliper the diameter of the cutting edges. Eccentric fluting, as it is called, consists in milling the flutes with uneven spacing to obviate chattering, but having them exactly opposite, so that a diameter measurement may be made with a micrometer.

A rose-reamer is an end-cutting tool, and is often used in place of a drill in cored holes. It is never made for hand use, and in general practice is seldom used for exact diameter.

MILLING CUTTERS. In lathe work the cutting tool is fixed and the work rotates. In a milling machine the cutter rotates and work is fed against it. The rotating cutter, termed a "milling cutter," has an almost unlimited variety of sizes and shapes for milling regular and irregular forms. Milling cutters are made from some of the tool steels, heat-treated to give the right cutting qualities, the stock coming to the tool-maker in the form of rough blanks, carefully annealed. Where the cutter has a hole through it this is first drilled, bored, or reamed to a diameter somewhat smaller than that in the finished cutter. The reason for this is that all the exact true surfaces must be finished after the cutter has been hardened — some grinding process being necessary which requires an excess of stock.

When the length of the cutter is greater than about one-half inch, it is usual to chamber the hole to a shape that renders it necessary to diameter grind the holes at the ends only. In cutters of considerable length the saving in grinding by this procedure is considerable. The sides of the blanks are usually recessed, giving a hub-and-rim effect at the sides of the cutter. An even number of teeth is preferable, and these are spaced to a circumferential pitch varying from three-eighths to three-quarters inch for ordinary cutter sizes.

When the teeth are milled into the solid blank, a cutter giving a space angle of sixty degrees is preferred for cutting the peripheral teeth, while one of seventy degrees is generally used for the side teeth. Where milling cutters are made in quantity, special space cutters are worked out to give the maximum chip room consistent with tooth strength.

After the cutter has been heat-treated to the proper hardness, it is finished to the specific dimensions by grinding.

GRINDING THE HOLE. Unless special methods and tools are employed the hole is completely finished as the first operation of grinding. This is accomplished by holding the cutter trued in a chuck screwed on the spindle of a Universal grinder and grinding out the hole to standard size, using an internal grinding attachment.

GRINDING THE SIDES. Fig. 30 shows how to grind the sides with the cutter held flat against a faceplate. If the cutter is to be used for deep cuts, the faceplate is set to give a slight concavity to the sides of the cutter.

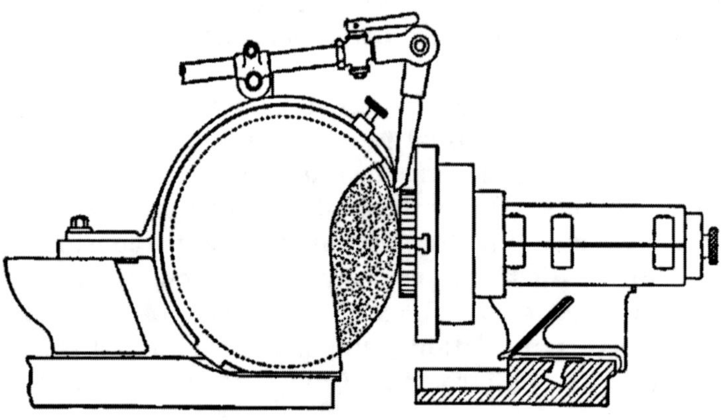

FIG. 30

CLEARANCE OF THE TEETH. The teeth of milling cutters are given a slight clearance back from the cutting edges; five degrees is usually sufficient.

JIGS AND FIXTURES

Jigs and fixtures are special devices designed to put manufacturing upon an efficient basis. Three distinct purposes are served by the use of jigs: (*a*) Reduction of cost per piece; (*b*) interchangeability of parts; and (*c*) accurate production.

Jigs and fixtures are usually made from cast iron or steel. Their use practically does away with fitting, as this term is known in shops not using jigs.

JIG DESIGN. A jig is a device for holding the work and for locating the tool work to be done upon it. A good example of this is shown in the drill jig, Fig. 31.

Jigs are of the plate type which lies upon and is clamped to the surface of the work; of the open-box type; and of the closed-box type.

In designing a jig, the piece is first drawn upon a sheet of paper, which is sufficiently large to allow locating the views some distance apart. This permits building the jig in the drawing around the "coupon," as the piece is often called. To start the design, first determine and lay down the locating points or stops, then arrange the clamping device. A jig should be so designed that the work can be put into position in only one way. Provide for supporting the thrust of the cutting tools in such a manner as to avoid springing the work. Make the jig as simple as possible, avoiding every feature in design that complicates the workman's use.

While in the larger shops the jigs are designed by the draftsmen, in many shops the tool-maker both designs and builds the jigs, and in no other way can a workman so clearly show his ability and ingenuity as in the building of jigs.

JIG BODY. The jig body is usually of cast iron, which is first rough planed or milled on all surfaces which are to be finished. These surfaces are then finish planed to final dimensions. In some cases jig bodies are finished by grinding in a surface grinder.

LOCATING BUSHING HOLES. If no particular accuracy is demanded, the holes for bushings can be located directly by careful attention to ordinary laying-out methods, and the hole drilled and reamed directly.

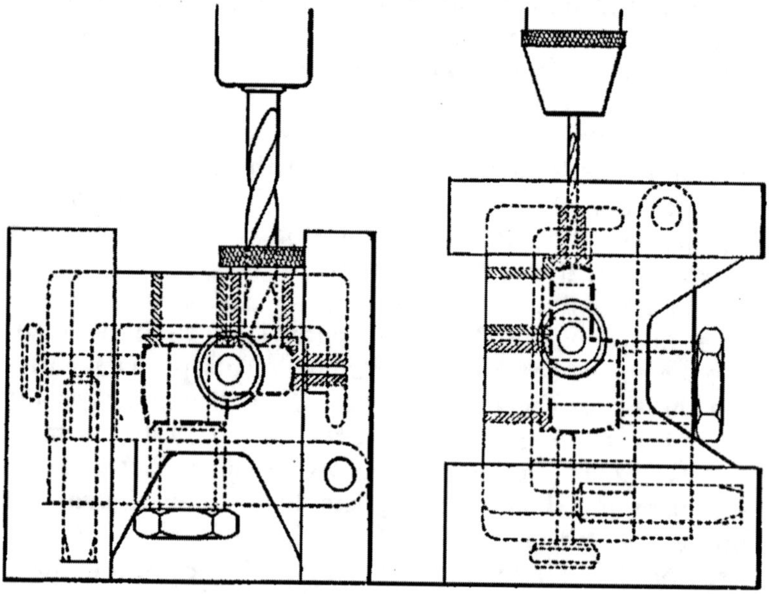

FIG. 31

When the allowable error is very small a more accurate scheme must be followed, and the best of several methods for the average tool-maker is that known as the button method. In this the holes are located by laying out scribed center lines and locating intersections where the holes are to be centered. Instead of drilling and reaming the bushing holes, holes are drilled and tapped to fit the button screws. The jig buttons are small, accurately ground cylinders, as shown in Fig. 32. These are held by means of the screws, lightly clamped in place, and exactly located to centers by accurate measurements. The highest possible accuracy in locating holes is secured by this method.

FIG. 32

BORING HOLES. The holes for the hardened bushings are usually bored by swinging the jig body upon a face-plate in an engine lathe. The jig body is then shifted upon the face-plate until a button indicates true

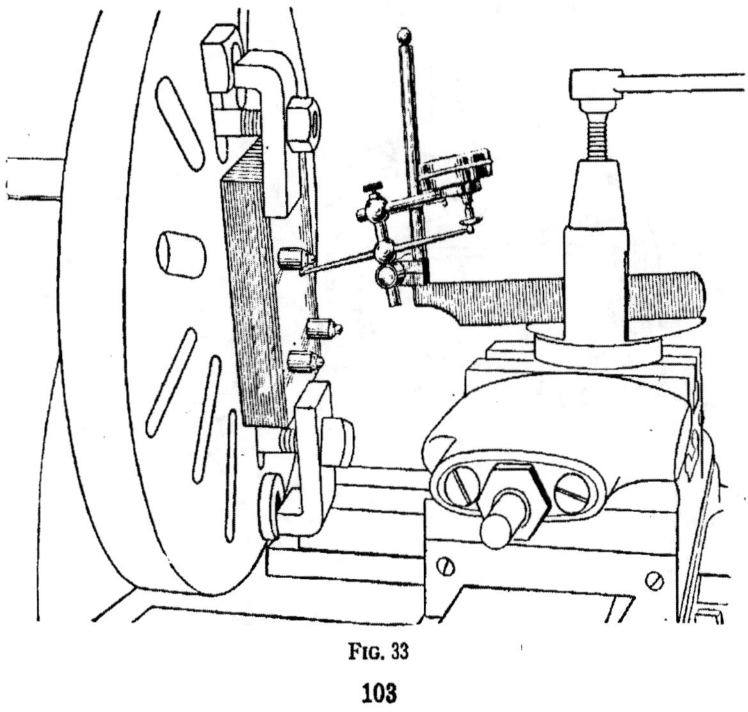

FIG. 33

with a Universal Dial Indicator, as shown in Fig. 33. The jig body is then clamped tightly upon the face-plate. After removing the jig button, the hole is first rough-

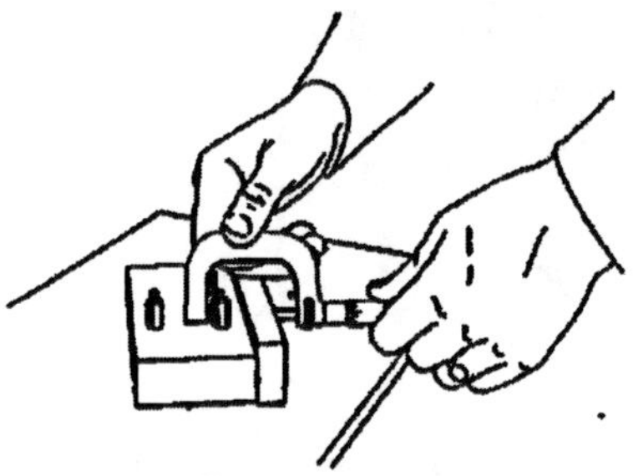

ADJUSTING BUTTONS TO SIDE OF PLATE

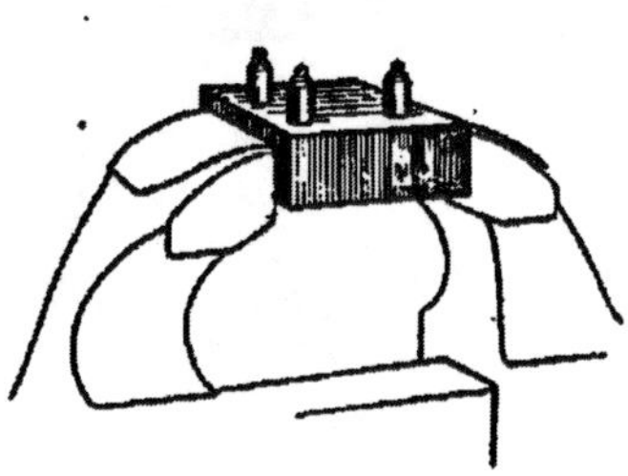

BUTTONS IN PLACE

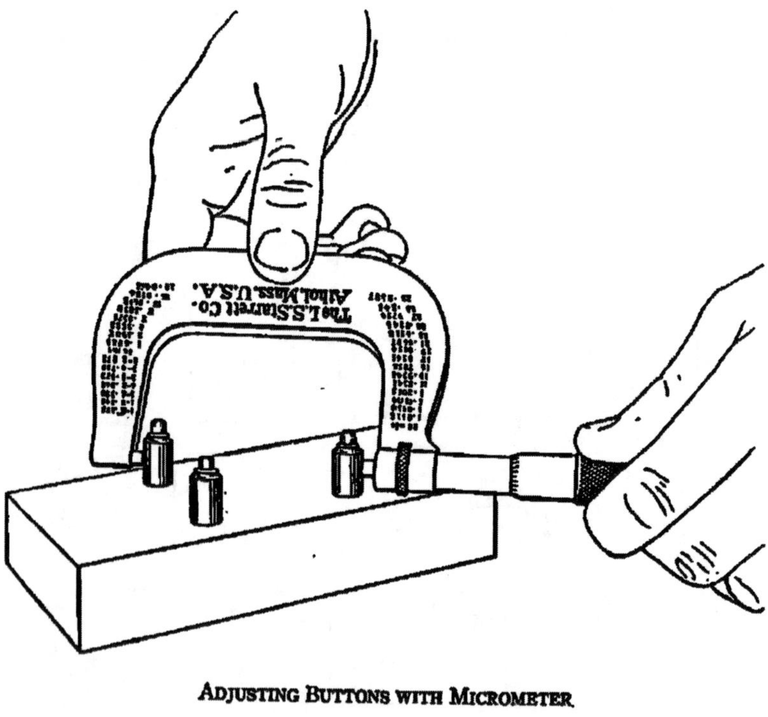

ADJUSTING BUTTONS WITH MICROMETER

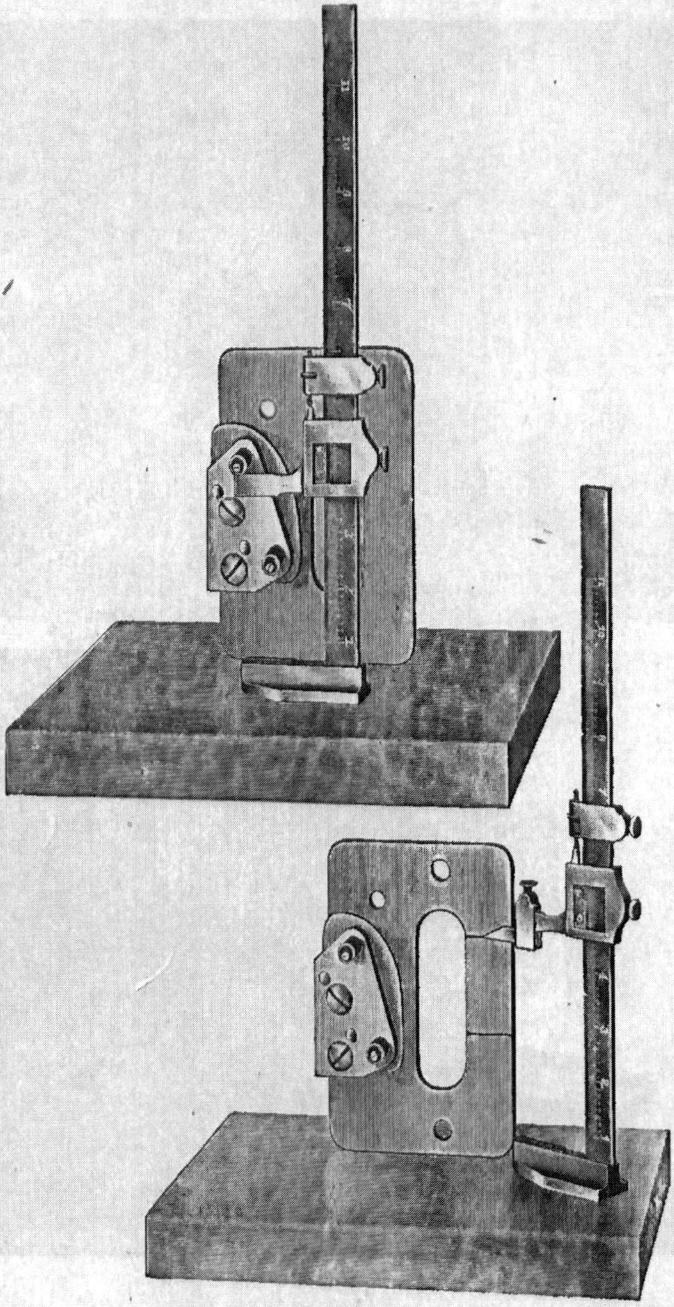

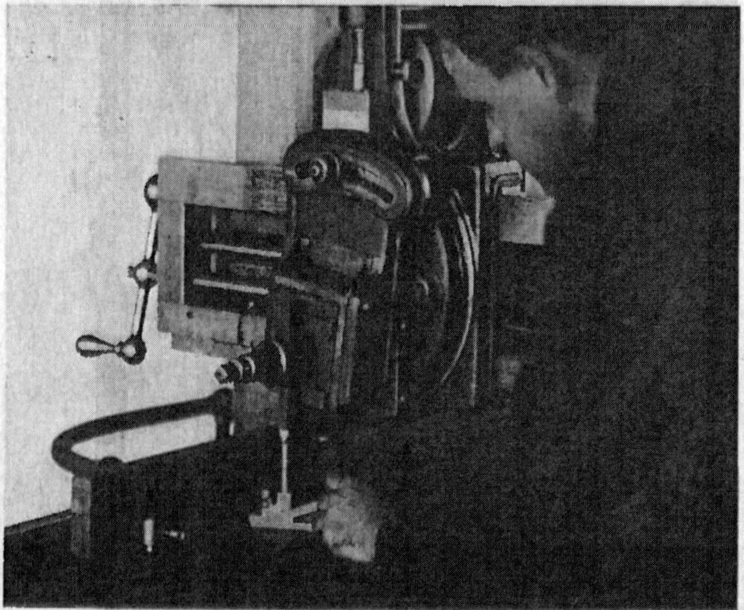

drilled approximately to size, and afterwards carefully bored exactly to size. This prepares the hole for holding the hardened steel bushing; the process is repeated for all the previously located buttons.

JIG BUSHINGS. If the holes in a cast-iron or soft-steel jig body were left as bored, they would soon lose accuracy by wearing off center. To prevent this wear the holes are lined with hardened and carefully ground bushings, pressed or driven tightly into place. These

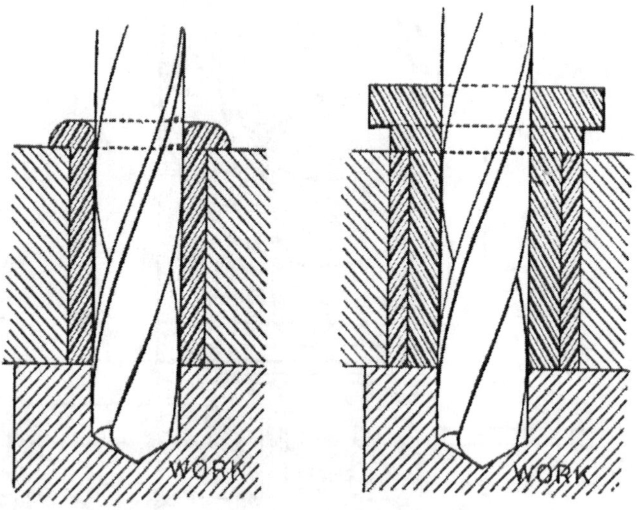

bushings are made with a hole having a diameter equal to that of the tool which passes through them. The bushings are sufficiently long to support the drill. In case the jig bushings must be removed frequently, they are known as slip bushings, and the hole in which they slip is lined with a steel lining, itself hardened and ground. In some cases the bushing locates the work as well as the tool, and if so the bushing screws through the body of the jig and against some prominent part of the work, as a boss for example.

TOLERANCES. In all construction work a certain amount of inexactness is allowable. In other words, it is impossible to obtain absolute precision, and the allowable errors in exactness are termed "tolerances." In some cases a tolerance of one-sixteenth inch might be allowed, while in others exactness to the fraction of a thousandth part of an inch may be necessary. See pages 31 and 32.

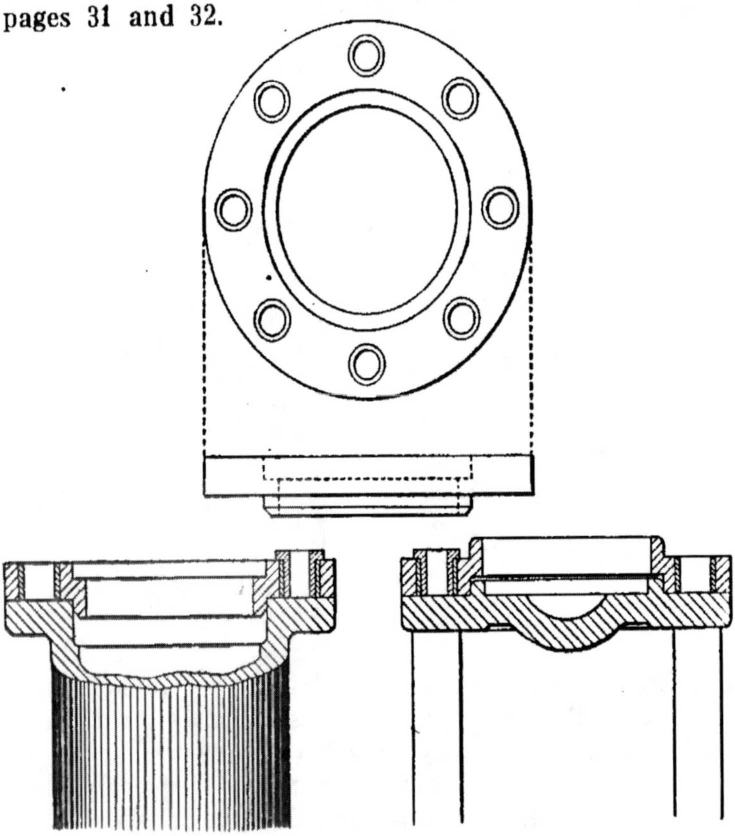

JIG FOR DRILLING BOLT HOLES IN CYLINDER FLANGE AND HEAD

The projection on the jig keeps it concentric with the bore of the cylinder, and the recess fits over the projection on the head.

GRINDING

In the machine shop the term "grinding" refers to the producing of finished surfaces by means of rotating grinding wheels, and the process of grinding as used in finishing machine parts is to-day the most efficient method devised for the purpose. With a proper selection of grinding machine and grinding wheel, all of the common machine construction materials may be readily and accurately finished.

Grinding machines are classified into two groups, (*a*) those for curved surfaces; as, for example, cylindrical work; and (*b*) those for plane or flat surfaces. The first of these is usually called a cylindrical grinder, and the second is known as a surface grinder. Each group has many designs, made necessary by the varied uses to which grinding is adapting itself.

GRINDING WHEELS. These are now known as abrasive wheels, and the material from which they are made is termed an abrasive. The abrasives in common use are the minerals emery and corundum, and the manufactured abrasives, sold under the trade names of Alundum, Aloxite, Carborundum, Crystolon. Owing to the uniformity of the product as it comes from the electric furnace, manufactured abrasives are at present more largely used than natural abrasives.

MAKING ABRASIVE WHEELS. An abrasive wheel is made up of one of the above-named ABRASIVES and a BOND. The bond is, as its name indicates, something for holding the abrasive in mixture. Grinding wheels are made by three processes, known as Vitrified, Silicate, and Elastic.

VITRIFIED WHEELS. In wheels made by the Vitrified process, the bond is of earth or clay which hardens or vitrifies when subjected to a temperature of about 2500° F. to 2800° F. for a definite period of time. Vari-

Allowances for Grinding—Table 10

Diameter, Inches	Length, Inches										
	3	6	9	12	15	18	24	30	36	42	48
	Allowance, Inches										
½	0.010	0.010	0.010	0.010	0.015	0.015	0.015	0.020	0.020	0.020	0.020
¾	0.010	0.010	0.010	0.010	0.015	0.015	0.015	0.020	0.020	0.020	0.020
1	0.010	0.010	0.010	0.015	0.015	0.015	0.015	0.020	0.020	0.020	0.020
1¼	0.010	0.010	0.015	0.015	0.015	0.015	0.015	0.020	0.020	0.020	0.020
1½	0.010	0.015	0.015	0.015	0.015	0.015	0.020	0.020	0.020	0.020	0.020
2	0.015	0.015	0.015	0.015	0.015	0.020	0.020	0.020	0.020	0.020	0.025
2¼	0.015	0.015	0.015	0.015	0.020	0.020	0.020	0.020	0.020	0.025	0.025
2½	0.015	0.015	0.015	0.020	0.020	0.020	0.020	0.020	0.025	0.025	0.025
3	0.015	0.015	0.020	0.020	0.020	0.020	0.020	0.025	0.025	0.025	0.025
3½	0.015	0.020	0.020	0.020	0.020	0.020	0.025	0.025	0.025	0.025	0.025
4	0.020	0.020	0.020	0.020	0.020	0.025	0.025	0.025	0.025	0.025	0.030
4½	0.020	0.020	0.020	0.020	0.025	0.025	0.025	0.025	0.025	0.030	0.030
5	0.020	0.020	0.020	0.025	0.025	0.025	0.025	0.025	0.030	0.030	0.030
6	0.020	0.020	0.025	0.025	0.025	0.025	0.025	0.030	0.030	0.030	0.030
7	0.020	0.025	0.025	0.025	0.025	0.025	0.030	0.030	0.030	0.030	0.030
8	0.025	0.025	0.025	0.025	0.025	0.030	0.030	0.030	0.300	0.030	0.030
9	0.025	0.025	0.025	0.025	0.030	0.030	0.030	0.030	0.300	0.030	0.030
10	0.025	0.025	0.025	0.030	0.030	0.030	0.030	0.030	0.030	0.030	0.030
11	0.025	0.025	0.030	0.030	0.030	0.030	0.030	0.030	0.030	0.030	0.030
12	0.030	0.030	0.030	0.030	0.030	0.030	0.030	0.030	0.030	0.030	0.030

ous grades of hardness are obtained by using bonds of different tensile strength. The ideal bond is one which retains the grains of abrasive until sufficiently dulled by use, and then allows them to break away, and in this manner bring fresh cutting edges and points into grinding contact.

SILICATE WHEELS. Silicate of Soda is the bond used in silicate wheels; and wheels made by this process are most efficient for tool and knife grinding.

ELASTIC WHEELS. This process of bonding is generally used for the very thin wheels used for slitting metals. The principal ingredient of the bond is shellac.

GRADING THE ABRASIVE. By numerous crushing, grinding, cleansing, and sorting processes, the abrasive is graded into a series of sizes which give the wheel its grain number. This number conforms to the sieve mesh through which the abrasive is passed; for example, grain No. 40 indicates that the abrasive was graded through a sieve having a mesh of forty to the linear inch.

COMBINATION WHEELS. For many grinding purposes the combination wheel is preferred to a wheel of single grade. Combination wheels are made up of abrasives of several grain numbers.

BONDING. The ideal bond is one which is impervious to moisture, does not soften by heat, and which holds firmly the cutting points of the abrasive until they become dulled by use. The bond then releases the dull abrasive and permits fresh, sharp points to begin cutting. With abrasives of equal quality the maker who nearest approaches the ideal bond produces the superior wheel.

GRADING THE WHEELS. In grinders' language, abrasive wheels are known as hard wheels and soft wheels. The maker, therefore, lists his wheels as hard or soft by some scale of numbers or by letters. A prominent firm uses the letters of the alphabet, as shown in the following list in which "M" is medium.

Norton Grade List

The following grade list is used to designate the degree of hardness of our Vitrified and Silicate Wheels, both Alundum and Crystolon.

E....................................Soft
F
G
H
I....................Medium Soft
J
K
L
MEDIUM..........M................MEDIUM
N
O
P
Medium Hard...........Q
R
S
T
Hard.........................U
V
W
X
Extremely Hard.....................Y
Z

The intermediate letters between those designated as soft, medium soft, etc., indicate so many degrees harder or softer; *e. g.*, L is one grade or degree softer than medium; O, two degrees harder than medium, but not quite medium hard.

Elastic Wheels are graded as follows: 1, 1½, 2, 2½, 3, 4, 5, and 6. Grade 1 is the softest and grade 6 the hardest.

CYLINDRICAL GRINDING. When the piece being ground is rotated, the process is known as cylindrical grinding, and the development of machines for grinding cylinders has given the process a great impetus. While it is possible to grind from the rough stock without previous lathe work, the method usually followed is to first rough turn the work.

ROUGHING FOR GRINDING. This process includes the work done in removing excess stock previous to finishing to size in the grinding machine. Unless a study is made of the conditions surrounding the whole operations of the lathe and the grinding machine, lack of efficiency may result. In general where the work is to be ground it is best to consider the lathe as a mere roughing machine for removing the excess of stock at as deep a cut and as coarse a feed as is consistent with an efficient cutting speed, leaving the job of finishing to the grinding machine.

AMOUNT TO LEAVE FOR GRINDING. If the grinding machine is modern in design as much as 1/32 of an inch, or even more may be left on machinery steel parts for removal in the grinder; the amount varying with the size of the work itself. An allowance of 1/64 of an inch is general on the smaller machine parts, but this allowance should be increased on larger sizes. Table 10, page 110, shows allowance for grinding as recommended by one maker of grinding machines, and Table 11 shows grinding wheel speeds.

SELECTING THE WHEEL. The selection of the wheel to be used in any grinding operation can, perhaps, best be made by reference to Table 12, page 115, which fairly represents general practice. As the hardness of material and the area of contact made by the wheel have a marked influence, no table can entirely solve the problem, but it may be used as a start in the right direction. In general a soft wheel should be used on hardened work and a harder wheel on soft materials.

Table of Grinding Wheel Speeds—Table 11

Diameter Wheel	Millimeters	Rev. per Minute for Surface Speed of 4,000 Feet, or 1,200 Meters	Rev. per Minute for Surface Speed of 5,000 Feet, or 1,500 Meters	Rev. per Minute for Surface Speed of 5,000 Feet, or 1,800 Meters
1 inch	about 25	15,279	19,099	22,918
2 "	" 50	7,639	9,549	11,459
3 "	" 75	5,093	6,366	7,639
4 "	" 100	3,820	4,775	5,730
5 "	" 125	3,056	3,820	4,584
6 "	" 150	2,546	3,183	3,820
7 "	" 175	2,183	2,728	3,274
8 "	" 200	1,910	2,387	2,865
10 "	" 250	1,528	1,910	2,292
12 "	" 305	1,273	1,592	1,910
14 "	" 355	1,091	1,364	1,637
16 "	" 405	955	1,194	1,432
18 "	" 455	849	1,061	1,273
20 "	" 505	764	955	1,146
22 "	" 515	694	868	1,042
24 "	" 610	637	796	955
26 "	" 660	586	733	879
28 "	" 710	546	683	819
30 "	" 760	509	637	764
32 "	" 810	477	596	716
34 "	" 860	449	561	674
36 "	" 910	424	531	637
38 "	" 965	402	503	603
40 "	" 1,015	382	478	573
42 "	" 1,065	364	455	546
44 "	" 1,115	347	434	521
46 "	" 1,165	332	415	498
48 "	" 1,220	318	397	477
50 "	" 1,270	306	383	459
52 "	" 1,320	294	369	441
54 "	" 1,370	283	354	425
56 "	" 1,420	273	341	410
58 "	" 1,470	264	330	396
60 "	" 1,520	255	319	383

The R. P. M. at which wheels are run is dependent on conditions and style of machine and the work to be ground.

Wheels are run in actual practice from 4,000 to 6,000 feet per minute; in some instances as high as 7,500 feet.

Grade and Grain of Grinding Wheels for Different Materials*

Table 12

(The Norton Co.)

Class of Work	Alundum		Crystolon	
	Grain	Grade	Grain	Grade
Aluminum castings	36 to 46	3 to 4 Elas.	20 to 24	P to R
Brass or bronze castings (large)	....	....	20 to 24	Q to R
Brass or bronze castings (small)	....	....	24 to 36	P to R
Car wheels, cast iron	....	....	16 to 24	P to R
Car wheels, chilled	20	Q	16 to 24	O to Q
Cast iron, cylindrical	24 comb.	J to K	30 to 46	J to L
Cast iron, surfacing	20 to 46	H to K	16 to 30	J to L
Cast-iron (small) castings	24 to 30	P to R	20 to 30	Q to S
Cast-iron (large) castings	16 to 20	Q to R	16 to 24	Q to S
Chilled iron castings	20 to 30	P to U	20 to 30	Q
Dies, chilled iron	....	....	20 to 30	O to Q
Dies, steel	36 to 60	J to L	....	....
Drop-forgings	20 to 30	P to R	....	....
Internal cylinder grinding	....	....	30 to 60	I to L
Internal grinding, hardened steel	46 to 60	J to M	....	....
Machine shop use, general	20 to 36	O to Q	....	....
Malleable iron castings (large)	14 to 20	P to U	16 to 20	R to S
Malleable iron castings (small)	20 to 30	P to R	20 to 30	Q to S
Milling cutters, machine grinding	46 to 60	H to M	....	....
Milling cutters, hand grinding	46 to 60	J to M	....	....
Nickel castings	20 to 24	P to Q	20 to 25	R
Pulleys, surfacing cast iron	....	....	30 to 36	K to L
Reamers, taps, etc., hand grinding	46 to 60	K to O	....	....
Reamers, taps, special machines	46 to 60	J to M	....	....
Rolls (cast iron), wet	24 to 36	J to M	24 to 36	J to M
Rolls (chilled iron), finishing	70	1½ to 2 Elas.	70 to 80	1½ to 2 Elas.
Rolls (chilled iron), roughing	....	....	30 to 46	2 to 3 Elas.
Rubber	30 to 50	J to K	30 to 50	K to M
Saws, gumming and sharpening	36 to 50	M to N	....	....
Saws, cold cutting-off	60	O to Q	....	....
Steel (soft), cylindrical grinding	24 comb.	L to N	....	....
	46 to 60	L to N	....	....
Steel (soft), surface grinding	24 to 36	H to K	....	....
Steel (hardened), cylindrical grinding	24 comb.	K	....	....
	46 to 60	J to L	....	....
Steel (hardened), surface grinding	36 to 46	H to K	....	....
Steel, large castings	12 to 20	Q to U	....	....
Steel, small castings	20 to 30	P to R	....	....
Steel (manganese), safe work	16 to 46	L to P	....	....
Structural steel	16 to 24	P to R	....	....
Twist drills, hand grinding	46 to 60	M	....	....
Twist drills, special machines	36 to 60	K to M	....	....
Wrought iron	12 to 30	P to U	....	....
Woodworking tools	46 to 60	K to M	....	....

* The information contained in this table is general and intended only to give an approximate idea of the grade used under ordinary conditions.

MOUNTING THE WHEEL. The wheel should be so mounted that there are no unequal stresses set up. Suitable guards should be provided to prevent injury to the workmen in case of the wheel bursting. The accompanying illustrations show RIGHT and WRONG methods of mounting wheels — carefully study the cuts.

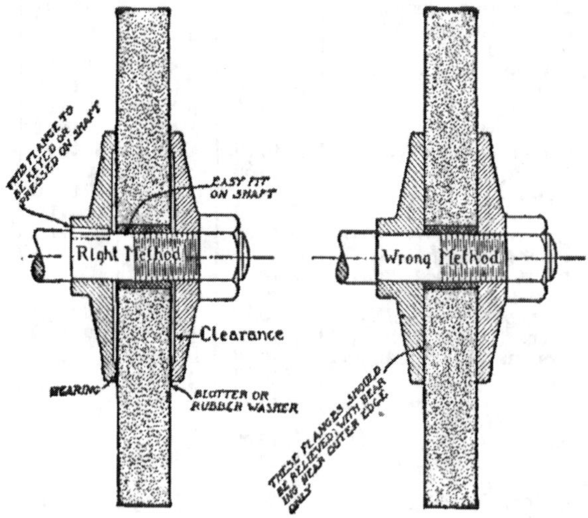

MEASURING THE WORK. The use of micrometers for obtaining exact measurements is nowhere better illustrated than in grinding. Fig. 34 shows an operator adjusting his micrometer for obtaining a measurement on a cylindrical piece, and Fig. 35 shows the operator as he makes his reading. While in lathe work the position of the operator leads naturally to adjusting the micrometer spindle with the fingers of the right hand, the left hand grasping the frame, in grinder work the reverse is generally true, hence he occupies the position as shown.

GRINDING FLAT SURFACES. Flat surface grinding may be divided into two general classes: (*a*) Machine

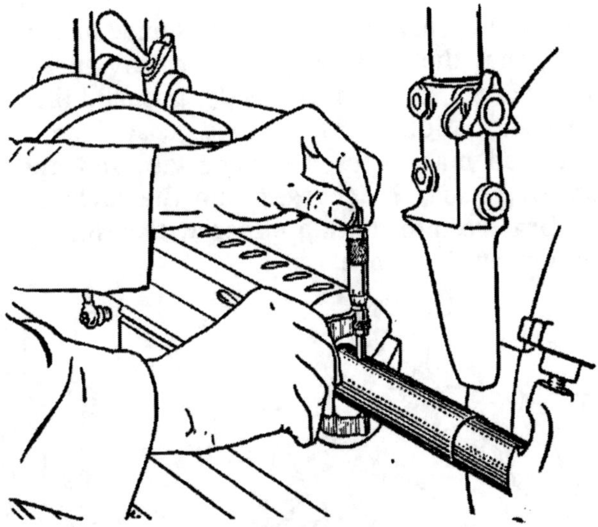

FIG. 34

parts, such as boxes, tables, cross-slides, faces of nuts, etc.; and (*b*) fine tool work, as, for example, steel blades, scales and rulers, straight edges, etc. Until recently the first-named class of work was done by reciprocating the work beneath the circumferential face of an abrasive wheel in a machine which, in principle, is not unlike a small planer. The use of machines with CUP WHEELS has practically revolutionized such grinding, and an exactness of surface is being obtained on fine flat work which leaves little to be desired.

LAPPING. In certain lines of work the final grinding process is often made, not with abrasive wheels as previously described, but by using metal discs, rings, or cylinders, the surfaces of which have been charged with a fine flour abrasive. Such a tool is called a "lap," and its use "lapping." Laps were first used by lapidaries in finishing the surfaces of mineral specimens, but laps have been in common use for a considerable time on fine work in the machine shop.

Laps are generally made of some material soft enough so that the abrasive can be readily pressed into the surface; or, as it is correctly termed, the surface is "charged." Soft, close-grained cast iron, copper, brass, or lead may be used for the lap, and any of the flour abrasives may be charged into the surface by rolling the abrasive into the lap either with a hardened roll or on a hardened surface.

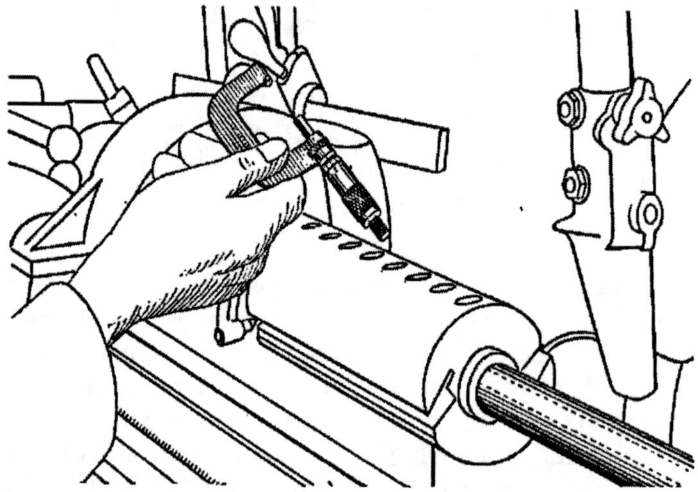

FIG. 35

In some of the finer grinding operations the lap is charged with diamond dust which has been precipitated or settled in a suitable dish of olive oil. The several grades are denoted by the time taken to precipitate; as, for example, fineness No. 5 takes ten hours.

Since lapping is a somewhat slow and tedious process it should be used only for the removal of small amounts of stock.

COMMON USES OF LAPPING. The more common uses of lapping are those of finishing micrometer ends, plug and ring gages, holes in jig bushings, and in the finest die and punch work.

LOCATING AND ALIGNING MACHINERY

When the product of the shop is determined, the proper location of the machines may be found by means of a plan or location drawing worked out in the drafting room. An easy way to do this is to provide rectangular slips of cardboard, each representing to some definite scale the plan outline of each machine. Placing these upon the floor plan of the room, the better of several arrangements may be found, and by using push pins the cardboard representations may be fixed in position.

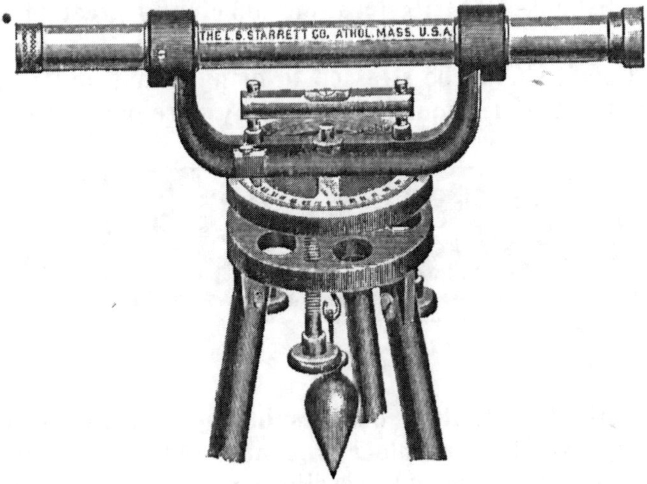

FIG. 36

Having decided upon the location, the machinery may be aligned in these positions by measurements from some base line made upon the floor or ceiling; or a leveling instrument,* such as shown in Fig. 36, may be used.

Ordinarily the machines are aligned by simple meas-

* See page 124 for directions for setting up a level.

urements and the countershafting hung from the ceiling vertically over the machine by plumbing up from the previously located machines. In such work thought must always be given to the line shafting and pulleys.

Unless care is used, there may be such interferences as to necessitate repeating the work. As the efficiency of the shop depends to a considerable extent on a convenient arrangement of the machines, all interferences should be taken care of on the ceiling rather than altering the arrangement of the machines.

ALIGNING THE SHAFTING. With the locations of the several lines of shafting determined upon, the usual method of alignment is to stretch a wire or cord the length of the room at the desired level of the shaft and at a distance from its location sufficiently great to give easy working room. With the two ends of the wire in position it should be stressed to bring it taut and should be supported at frequent intervals by wire hangers.

FIG. 37

With the shafting hangers in approximate position and the shafting in place, the necessary shifts can be made to bring the shaft parallel with the wire. A light stick notched at one end to rest upon the shaft and a wire brad at the other end for a feeler is all that is necessary for ordinary alignment. Leveling the shaft is done with special spirit levels having metal frames, the bases of which have been carefully grooved to set upon the shaft. Such a level is shown in Fig. 37. Special leveling and aligning attachments for setting and lining up shafting are sometimes used. Shafting is often lined by plumbing up from a data line on the shop floor with a mercury plumb bob.

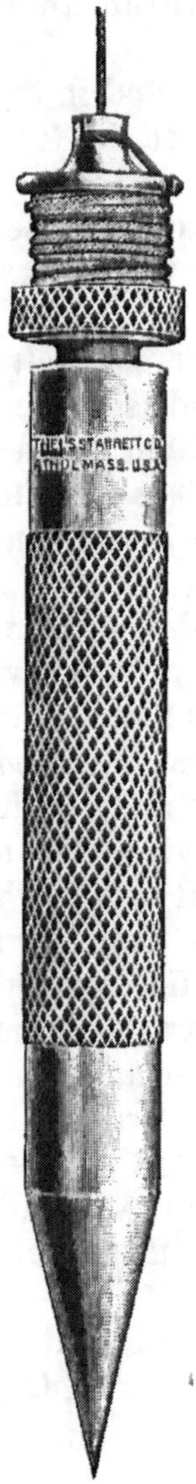

Mercury Plumb Bobs

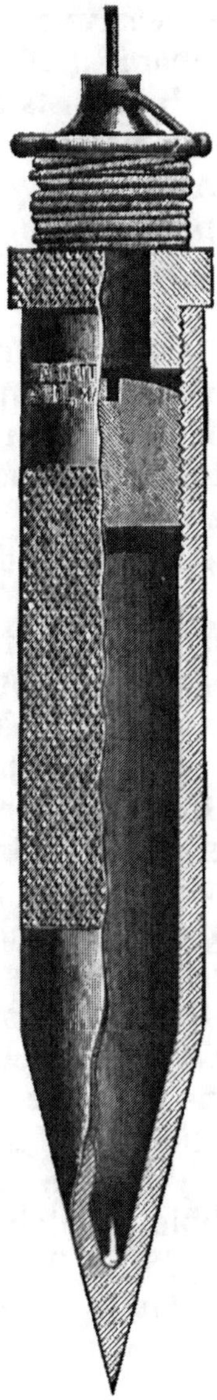

LEVELING INSTRUMENT

While the surveyors' transit can be used in shop leveling and in shaft aligning a much simpler and a more inexpensive instrument termed a leveling instrument is all that is needed.

It consists of a table capable of being adjusted in the horizontal plane, which carries a yoke which in turn carries a twelve-inch brass tube. The whole instrument is placed upon a suitable tripod. The tube has no lenses and therefore is not a telescope as in the surveyors' instrument.

At one end of the tube are the usual cross hairs which locate the axis and at the opposite end is a peep hole or sight piece for the eye. The yoke which carries the tube is attached to a graduated arc which is let into the upper part of the table; this allows the instrument to swing to read angles in the horizontal plane.

ADJUSTING THE INSTRUMENT. In using this instrument it is important that the table be carefully leveled. It is pivoted on the tripod tube by a ball and socket joint. Three knurled-head adjusting screws threaded through the tripod top and resting against the under side of the table furnish a means of adjusting the table. Upon the table carrying the yoke is a bent-tube spirit level with a sensitive air bubble. After the tripod legs have been placed to roughly level the instrument, adjust the knurled leveling screws to give as correct a centering for the air bubble as is possible. To test this adjustment swing the yoke, which carries the air bubble, to several positions and note any change in the position of the bubble. If there is a change, readjust the leveling screws until the yoke can be swung through its travel with the air bubble maintaining its central position.

USING THE LEVELING INSTRUMENT. While it is possible to so mount the leveling instrument upon a platform that its height will be sufficient for the use of targets mounted upon the shaft, the usual method is to hang targets upon the shaft and adjust them to swing low enough to allow the leveling instrument to be set with its tripod on the floor or on some convenient foundation spot.

THE TARGETS. These consist of stirrups which carry a spirit level and block with vertical and horizontal lines crossing each other. A plumb is hung upon the stirrup in such manner as to be readily raised or lowered. One of the targets may be hung upon the shaft free to swing plumb, the other is used as a fixed wall target.

USE. After the shafting has been roughly aligned with the wall of the building or with a line of columns, this being done by measurement, the leveling instrument is placed vertically beneath one end of the shaft. To locate the leveling instrument, plumb down from the center of the shaft, using the hanging target plumb bob, and locate a point in the floor or board placed on the foundation. A prick punch mark in the flat head of a wire brad previously driven into the floor provides a permanent point. Set the tripod of the leveling instrument directly over this point, using the plumb bob hanging from the center of the table. Next carefully level the table as already described. Hang the portable target closely in front of the cross-hair end of the tube and level and adjust its height until the horizontal cross hair of the tube coincides with the horizontal cross line of the target.

Remove the target to the far end of the shaft and swing the tube of the leveling instrument until the sight through the tube coincides with the vertical line on the target. With the hanging target displaced, mount a fixed target upon the wall at the far end of the shaft and adjust it until its cross lines coincide with the cross hairs of the tube as sighted. If the instrument is in its original position with the plumb bob over the point in the floor, the setting up of the instrument is complete. By reference to the fixed target it can at all times be checked.

Replace the hanging target at the far end of the shaft and adjust the adjacent hanger so that the cross lines of the target coincide with the cross hairs when sighting through the tube. Repeat for each hanger until the target can be hung upon the shaft adjacent to any hanger and show perfect coincidence of target cross lines and tube cross hairs.

Note that after the instrument and target have been set neither should receive further adjustment except in case of accident—the shaft itself receives the adjustments.

HOW TO SET UP THE TRANSIT

The Starrett transit or level can be used for the same purposes as any engineer's transit and level, and because of its simplicity and freedom from complications, it can be used by any one in laying out foundations for buildings, aligning machinery, and in building dams and raceways for simple water-power developments.

The transit combines in one instrument the facilities for measuring both horizontal and vertical angles, and enables the operator to lay out anything that does not require excessive refinement. The level is for measuring angles in a horizontal plane only, and it should be borne in mind that the level will do all that the transit will do, except measure vertical angles. The transit, which is furnished either with a telescope or plain-sight tube, is mounted on a tripod, and has a plate carrying a graduated arc. The telescope or sight-tube is connected to a graduated vertical arc so that vertical angles may be measured as well as horizontal. It is provided with leveling screws, and with a ground level vial for adjusting the level of the graduated plate.

To level the instrument, the legs must be firmly set into the ground or floor, so that neither wind nor accidental touch will disturb the adjustment. It should then be made as nearly level as possible by adjusting the lower parts of the extension legs. It should then be brought to a perfect level by means of the leveling screws between the plate and tripod head. This is done by bringing the level over any one of the leveling screws and turning one screw in and another out until the bubble appears in the center of the level glass. The sight tube or telescope should then be turned through an angle of about ninety degrees and again the bubble adjusted to the center of the glass by means of two leveling screws. This operation should be continued until the bubble stands in the center of the glass, no matter in what direction the telescope may be turned.

To find differences of level of two places, the instrument should be placed in a position about equally distant from the two points. First obtain the height of the target on one of the rods by means of the cross line in telescope or sight tube and make record of the same. Then carry the rod to the other position and find the height of the target at that point. The difference between the two heights, as read on the rod, will be the difference of level of the two places, that place being higher at which the height of the target is less.

ELEMENTARY ALGEBRA

Many engineering and shop problems can be solved more readily with algebra than by means of arithmetic. In fact, some problems cannot be solved by arithmetic; as, for example, when the conditions are not fully and concretely stated. Algebra is applied by expressing the relations in algebraic terms, forming them into an equation, which states the conditions, and then solving the equation.

In arithmetic a figure has a definite value, 4 or 20 for instance, and the value remains unchanged; it is always 4 or 20. In algebra letters are used, and as these letters do not always have a definite value, their use adds flexibility to mathematical operations. Some find it easier at the beginning to think of the letters as abbreviations.

SYMBOLS

Some of the symbols or signs of algebra are the same as those used in arithmetic.

THE SYMBOLS OF QUANTITY are the figures used in arithmetic and the letters of the alphabet.

THE COMMON SYMBOLS OF OPERATION are the signs used in arithmetic; they are as follows:

$+$ is the sign of addition, called plus. If no sign precedes numbers or letters the plus sign is understood; that is, $2abc$ is $+2abc$.

$-$ is the sign of subtraction, or difference, called minus.

$\times$ is the sign of multiplication, called times. When there is no sign between letters or between letters and figures, multiplication is understood. Thus $3cd$ means $3 \times c \times d$. But this does not apply to numbers: 328 is not $3 \times 2 \times 8$, but 328, same as in arithmetic.

$\div$ is the sign of division, read "divided by." Division may also be expressed by a horizontal line between the quantities, as, $a \div b = \frac{a}{b}$ or $\frac{16}{4} = 16 \div 4$.

COEFFICIENT. The numerical factor or number is generally called the coefficient; in $5abc$, 5 is the coefficient; but, strictly speaking, $5a$ is the coefficient of bc, and $5ab$ is the coefficient of c. Again in the expression $3a\ (b - c)$, $3a$ is the coefficient of $(b - c)$, or in the expression $(a + b)\ x$, $(a + b)$ is the coefficient of x.

When no numerical coefficient is expressed, it is always unity or 1. Thus $a = 1a$.

EXPONENT. The small figure or letter written at the right and a little above a number or letter is called the exponent; it shows how many times the number is to be taken as a factor.

Thus 2^2 is read "2 squared" or "2 with the exponent 2." The number 2 is to be used twice as a factor, or multiplied by itself. Similarly a^3 is read "a cubed" or "a with the exponent 3." The letter a is to be taken three times as a factor, or $a \times a \times a$. In the same way $(m + n)^4 = (m + n) \times (m + n) \times (m + n) \times (m + n)$.

Again $a^2bc^3d^4 = a \times a \times b \times c \times c \times c \times d \times d \times d \times d$.

Note this difference —

$$m^4 = m \times m \times m \times m$$
$$4m = m + m + m + m$$

SYMBOLS OF RELATION show the relative values of letters.

$=$ is the sign of equality, read "equals" or "equal to." $a = b$ means that a is equal to b, or whatever value is given to a, the same value must be given to b. If $4a = 3b$, 4 times some quantity represented by a is equal to 3 times some quantity represented by b, but it is evident that a does not equal b.

: is read "is to" or "to." It indicates ratio.

If two ratios are equal, they may, of course, be connected by the sign of equality, but more often they are connected by this sign ::

SYMBOLS OF AGGREGATION

( ) Parentheses.
[ ] Brackets.
{ } Braces.
—— Vinculum.
√ Radical Sign (square root).

Letters or quantities enclosed in parentheses are to be handled as a single quantity.

$5\ (c + d)$ means that $c + d$ as one quantity is to be multiplied by 5.

Or $(a + b) \div (x + y)$ means that $a + b$ taken as a single quantity is to be divided by $x + y$ taken as a single quantity. Another way of expressing it is, the same operation performed on a must be performed on b also.

Again $-(a + b)$ means that the sum of a and b taken as a single quantity is to be subtracted. It does not mean that a alone is to be subtracted.

THE RADICAL SIGN. This sign is used as in arithmetic; that is, it shows that some root of the quantity is to be found, or expressed.

The small number or index used in connection with the radical sign denotes what root is meant. Thus $\sqrt[3]{a}$ is read "the cube root of a." $\sqrt[5]{b}$ is read the fifth root of b." When no index figure is used the square root is understood. $\sqrt{x + y}$ = the square root of $x + y$.

When the horizontal line extends over the expression it means that the indicated root is to be found of the entire expression. $\sqrt{m + n}$ = "the square root of $m + n$."

Let $m = 36$ and $n = 64$.

$$\sqrt{m + n} = \sqrt{36 + 64} = \sqrt{100} = 10$$
$$\sqrt{m} + n = \sqrt{36} + 64 = 6 + 64 = 70$$
$$\sqrt{m} + \sqrt{n} = \sqrt{36} + \sqrt{64} = 6 + 8 = 14$$

POSITIVE AND NEGATIVE TERMS

A term or quantity preceded by the plus sign, or by no sign at all, is a positive term, and one preceded by the minus sign is a negative term. This applies whether the term is a simple one like $3a$ (a monomial) or $(x + y)$ (a binomial) or $(a^2 + 2ab + b^2)$ (a polynomial).

SIMILAR TERMS. When several terms have the same letters, but may differ in numerical coefficients, they are called similar terms. Thus $4ac$, $-5ac$, and $3ac$ are similar terms.

In arithmetic we say that $+5$ and -5 cancel; that is, if we have five units and subtract five units we get zero. Similarly in algebra $5a$ cancels $-5a$, or $-6a^2xy$ cancels $6a^2xy$.

ADDITION

Addition is finding the sum of two or more quantities.

Arithmetic	Algebra
4 apples	$4ab$
3 apples	$3ab$
10 apples	$10ab$
17 apples	$17ab$

When the terms are alike, we add them by adding the coefficients; when they are not alike the addition is expressed.

$6ac$ added to $15xy$

$$6ac + 15xy$$

If the terms have different signs they can be added by algebra.

$$-6ac \text{ added to } 18ac = 12ac$$

$$-6ac \text{ added to } 18xy = 18xy - 6ac$$

When there are several quantities which are alike, but the signs unlike, we add them by adding all the positive or plus terms, then subtract the sum of all the negative or minus terms. For instance,

$$\begin{array}{r} 5mn \\ -2mn \\ 15mn \\ 3mn \\ -6mn \\ \hline 15mn \end{array}$$

The positive terms in the above equal $+23mn$ and the negative terms equal $-8mn$, the result being $23mn - 8mn = 15mn$.

Had all the signs been changed, the answer would have been $-15mn$; for the sign prefixed to the answer is that of the greater sum.

SUBTRACTION

Subtraction in many ways is like addition; that is, like terms can be subtracted in the same way that they can be added, and unlike terms are subtracted by indicating the difference.

Subtraction is the process of finding the DIFFERENCE between two quantities.

In arithmetic the larger cannot be subtracted from the smaller, but in algebra this can be done by expressing the difference.

In arithmetic 11 cannot be subtracted from 4, but in algebra $7 - 11 = -4$; that is, 7 lacks 4 of being equal to 11. It is minus 4.

The difference (in number of units) between 8 and 2 is 6, whether it is $8-2$ or $2-8$. Whether the difference is -6 or $+6$ depends upon which number is being subtracted.

These few rules should be remembered.

Subtracting a + quantity is the same as adding a minus quantity.

Subtracting a − quantity is the same as adding a plus quantity.

The sum of a minus quantity and a plus quantity is the difference between the quantities, with the prefixed sign of the larger.

The difference between a plus quantity and a minus quantity is equal to the sum of the quantities.

MULTIPLICATION

Multiplication is a short method of addition; that is, if you add $4ac$ five times, the result is the same as multiplying $4ac$ by 5.

$$\begin{array}{r} 4ac \\ 4ac \\ 4ac \\ 4ac \\ 4ac \\ \hline 20ac \end{array} \qquad\qquad \begin{array}{r} 4ac \\ 5 \\ \hline 20ac \end{array}$$

Multiplication is a process of taking a given quantity as many times as indicated by a number or another quantity.

Multiplication differs from addition in that unlike quantities can be multiplied.

$5abx$ multiplied by $6axy = 30a^2bx^2y$

This simple example shows that to multiply we first multiply the coefficients, then annex the letters, multiplying them when alike by adding the exponents; for instance, $a \times a = a^2$, $x \times x = x^2$.

SIGNS. If both quantities are plus, the product is plus; if both are minus, the product is plus; if one is plus and the other minus, the product is minus.

Multiplying more complicated quantities, those consisting of two or more terms each, is illustrated by this example in arithmetic:

Multiply $4 + 3 + 2 - 1$ by 6

Instead of adding before multiplying let us multiply each number by 6:

$$\begin{array}{l} 4 + \ 3 + \ 2 - 1 \\ 6 \\ \hline 24 + 18 + 12 - 6 = 48 \end{array}$$

If we use letters also, we proceed in the same way: Multiply $4ac + 3ab + 2c - c$ by $6a$.

$$\begin{array}{l} 4ac + \ 3ab + \ 2c - c \\ 6a \\ \hline 24a^2c + 18a^2b + 12ac - 6ac \end{array}$$

Combining similar terms, $24a^2c + 18a^2b + 6ac$

Multiply $2a + 4b$ by $3a - 6b$

$$\begin{array}{lll} 2a + 4b & & \\ 3a - 6b & & \\ \hline 6a^2 + 12ab & & \\ \quad - 12ab & - 24b^2 & \\ \hline 6a^2 & - 24b^2 & \end{array}$$

The above example should be thoroughly understood, for it involves multiplication, addition, and cancellation of like terms.

If three quantities are to be multiplied, first multiply two of them, then multiply the product by the third.

DIVISION

Division is the process of finding how many times one quantity is contained in another.

In arithmetic dividing 20 by 4 is finding how many times 4 is contained in 20.

In algebra dividing $25a^2bc$ by $5ac$ is finding how many times $5ac$ will go in $25a^2bc$.

First divide the coefficient 25 by 5, then divide the letters by subtracting the exponents of the same letter, $a^2 \div a = a$ because $2 - 1 = 1$. When no similar letter is in the dividend, as in the case of b, there is no exponent to subtract, therefore we put the b in the quotient. In the case of the letter c, c goes in c once or 1.

$$\begin{array}{l} 5ac\ )\ 25a^2bc\ (\ 5ab \\ \qquad\quad 25a\ \ c \\ \qquad\quad \overline{\quad ab \quad} \end{array}$$

Another way to state this is to divide the terms into factors:

$$\frac{25 \times a \times a \times b \times c}{5ac} = 5ab$$

The 5 cancels 5 in the numerator, a cancels one a in the numerator and c cancels c. These cancel because the exponents become zero; for instance, $1 - 1 = 0$, and c with the exponent zero equals one or unity.

SIGNS. Since division is the converse of multiplication, the rules governing signs are practically the same:

When both divisor and dividend are + the quotient is +.

When both divisor and dividend are − the quotient is +.

When the divisor is + and the dividend is − the quotient is −.

When the divisor is − and the dividend is + the quotient is −.

The process of polynomials is merely an extension of the process of dividing monomials.

Example: Divide $40a^4 - 35a^3b + 8a^2b - 7ab^2$ by $8a^2 - 7ab$:

$$8a^2 - 7ab)\ 40a^4 - 35a^3b + 8a^2b - 7ab^2\ (5a^2 + b$$

$$\underline{40a^4 - 35a^3b}$$

$$8a^2b - 7ab^2$$

$$8a^2b - 7ab^2$$

EQUATIONS

AN EQUATION is an algebraic expression in which two or more terms or quantities are connected by the sign of equality. The two terms or expressions are called members or sides of the equation; the term on the left-hand side is called the first, and that on the right-hand side is called the second term.

The letter whose value is to be found is called the "unknown quantity," and it is usual to represent the unknown quantity by the letter (x).

To solve an equation is to find the value of the unknown quantity, either in terms of numbers or in terms of numbers and letters.

A very important fact to remember about equations is that if the same operation is performed on both sides of the equation the left-hand side will still be equal to the right-hand side.

The equation will continue to be an equation if

a. The same quantity is added to both sides.
b. The same quantity is subtracted from both sides.
c. Both sides are divided by the same quantity.
d. Both sides are multiplied by the same quantity.
e. Both sides are raised to the same power.
f. The same root of both sides is extracted.

This fact is made use of in solving an equation; for instance,

$$5x = 20$$

Dividing both sides by 5, we have

$$x = 4$$

Again, $1/5x = 20$

Multiplying both sides by 5, we have

$$5 \times 1/5x = 5 \times 20$$
$$x = 100$$

Before solving an equation it is usually easier to rewrite or rearrange the terms so that x with its coefficient will be alone on the left-hand side. Changing the terms from one side to the other is called "transposing." It is evident that in transposing the truth of the sign of equality must not be destroyed.

Bearing in mind the fact that if the same operation is performed on both sides of an equation the left-hand side remains equal to the right-hand side, we can transpose terms.

$$x - 2a = b$$

Adding $2a$ to both sides, we have

$$x - 2a + 2a = b + 2a$$

As $-2a$ cancels $+2a$, we have

$$x = b + 2a$$

We see from this that the $2a$ has been transposed from one side to the other, and that in transposing the only thing that happened to it was that its sign was changed.

Numerous examples would show this simple fact that to transpose a quantity from one side of an equation to the other, it is only necessary to write the quantity on the other side with its sign changed; plus changed to minus or minus to plus.

If the term containing x is a fraction, the denominator can be eliminated, so that x will be alone, by multiplying both sides of the equation by the denominator.

$$\frac{x}{c} = \frac{m^2 + n^2}{b} - \frac{n}{b}$$

First, combine the fractions on the right-hand side, because they have the same denominator, thus:

$$\frac{x}{c} = \frac{m^2 + n^2 - n}{b}$$

To get x alone on the left-hand side, multiply both sides by c.

$$x = \frac{c\ (m^2 + n^2 - n)}{b}$$

Suppose x is in the denominator instead of in the numerator.

$$\frac{6}{x} = \frac{a + b}{10c}$$

Multiplying both sides by x gives

$$6 = \frac{(a + b)\ x}{10c}$$

Now transpose all terms

$$\frac{(a+b)\,x}{10c} = 6$$

Or dividing both sides by $\frac{a+b}{10c}$, the coefficient of x, we have

$$\frac{(a+b)\,x}{10c} \times \frac{10c}{a+b} = \frac{6\,(10c)}{a+b}$$

$$x = \frac{60c}{a+b}$$

The short cut to the same result is to invert both sides.

$$\frac{6}{x} = \frac{a+b}{10c}$$

$$\frac{x}{6} = \frac{10c}{a+b}$$

Then multiplying both sides by 6,

$$x = \frac{60c}{a+b}$$

SHOP AND ENGINEERING FORMULAS

The letters which we have used are given a meaning in shop and engineering formulas by assigning to each a definite numerical value. The letters are connected by signs to represent the conditions.

In a certain shop one-fifth of the output is milling machines, two-thirds is lathes, and the rest is twenty-eight shapers. How many milling machines and lathes are produced?

If we let x represent the total number of machines, $\frac{x}{5}$ equals the number of milling machines and $\frac{2x}{3}$ equals the number of lathes. The total is equal to $\frac{x}{5}$ added to $\frac{2x}{3}$, and this sum is added to 28 to equal the unknown quantity x.

$$x = \frac{x}{5} + \frac{2x}{3} + 28$$

Multiplying both sides by 15, the common denominator, to eliminate the fractions, we have

$$15x = 3x + 10x + 420$$
$$15x = 13x + 420$$

Transposing

$$15x - 13x = 420$$
$$2x = 420$$
$$x = 210$$

$\frac{x}{5} = \frac{210}{5} = 42$ milling machines and $\frac{2x}{3} = \frac{420}{3} = 140$ lathes.

In designing, formulas are used, and these formulas are in the form of equations, the letters having definite values. Usually the values of all but one letter are known or assumed. The problem then is to find the numerical value of the unknown by substituting the known values. For instance, in designing keys some use this formula:

$$P = \frac{126{,}000 \times H.\,P.}{D\,N}$$

in which P = the total twisting moment on the shaft, H. P. = the horse-power transmitted, D = diameter of shaft in inches, and N = number of revolutions of the shaft per minute.

If 20 horse-power is transmitted at a rotative speed of 40 revolutions per minute and the shaft is 2 inches in diameter, the twisting moment is found by substituting the known values and solving for P.

$$P = \frac{126{,}000 \times 20}{2 \times 40}$$
$$= 31{,}500$$

In finding the thickness of the hub of a pulley, some designers use this formula:

$$T = .14\sqrt[3]{B\,D}$$

in which T = thickness of hub in inches,
B = width of face in inches,
D = diameter of pulley in inches.

If the face is 8 inches and the pulley 27 inches in diameter, we have

$$T = .14\sqrt[3]{8 \times 27}$$
$$= .14 \times 6$$
$$= .84 \text{ inch or } 7/8 \text{ inch}$$

MENSURATION

ANGLES. Of all the plane figures which the machinist has to deal with, the angle is the most important, and also the most troublesome. Examples of working to an angle are found in the setting of the compound rest when taper turning, setting the head of the milling machine for milling spiral flutes in twist drills or reamers, and in the cutting of bevel gears. In laying out work the machinist must understand the properties of angles and the use of the protractor, so that he may work to the angle that is wanted, not to some other angle.

An angle is sometimes defined as the difference in direction of two straight lines; another definition is: an angle is the space between two straight lines that meet, or would meet if produced. Angles are also used for measuring rotation or circular movement.

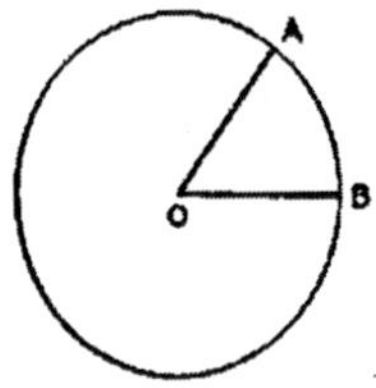

If a circumference of a circle is drawn, having for a center the vertex of the angle, the measure of the angle will be that arc included between the sides of the angle. Angle A O B is measured by the arc A B.

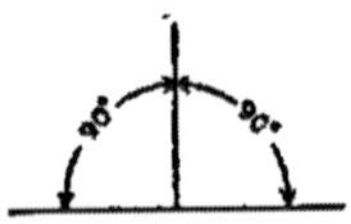

The circumference of the circle is divided into 360 equal parts, each called a degree. Each degree is divided into 60 equal parts called minutes. Each minute into 60 equal parts called seconds. The angle A O B will be an angle of 60° if the arc A B is one-sixth of the circumference.

It makes no difference what the radius of the circle or arc may be, the difference in direction is the same, and the number of degrees is the same.

A RIGHT ANGLE is one formed by two lines perpendicular to one another. The arc which measures it is a quarter circumference or 90°. The tool most commonly used for measuring a right angle is a try-square. Two right angles are formed when a line so meets another line that the two angles are equal.

AN ACUTE ANGLE is any angle of less than 90°.

AN OBTUSE ANGLE is any angle of more than 90°.

The complement of an angle is the angle which must be added to the given angle to make a right angle or 90°. The complement of an angle of 37° is 53°. Either of these angles is the complement of the other.

The supplement of an angle is the angle which must be added to the given angle to make 180°, or two right angles. The supplement of an angle of 63° is 117°. Either of these angles is the supplement of the other.

The instrument most commonly used for measuring angles is the protractor. It may be in the form of the combination set (page 14), or the protractor shown in the accompanying illustration. The protractor is a graduated disc on a fixed blade and adjustable stock. Any given angle may be laid out or measured by setting the blade at the desired angle with the stock. The angle shown here is a little less than 55°.

To set the protractor at an angle of less than 90° is an easy matter, because the instrument reads directly, being graduated from zero to 90°. But when the desired angle is greater than 90°, the supplement of the angle must be found and the protractor set to the supplement. Thus, to lay off an angle of 150° we first find the supplement or 30° and set the protractor at 30°. But the proper scale must be selected. It often happens that a protractor set to 60° actually measures 120°. With the Starrett combination set, all angles are read directly because of the two scales, each graduated from zero to 180°.

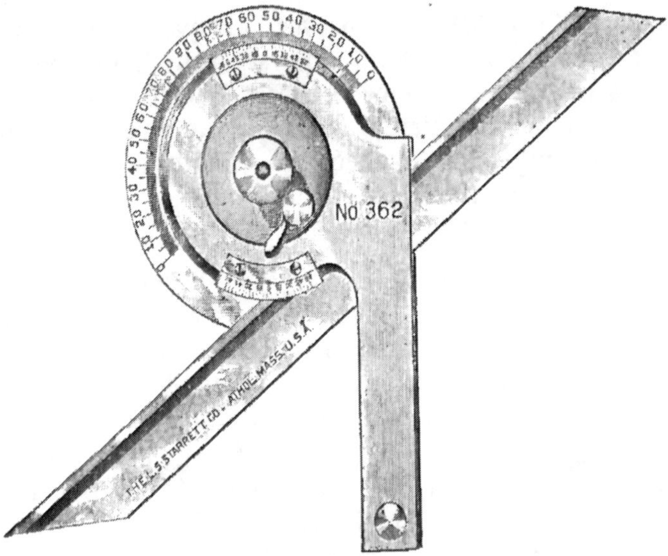

PROTRACTOR

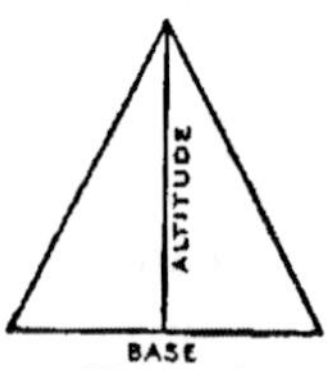

TRIANGLE

A plane figure of three sides — if all three sides are equal in length the triangle is equilateral and also equiangular; that is, all the angles are equal.

The sum of all three angles is equal to two right angles, or 180°.

Any angle equals 180° minus the sum of the other two.

The areas of two triangles are equal if they have equal base and equal height or altitude.

If the three sides of a triangle are proportional to the corresponding sides of another triangle, the triangles are *similar* and the corresponding angles are equal.

If the angles of a triangle are equal to the corresponding angles of another triangle, the triangles are *similar* and the corresponding sides are proportional.

The area of any triangle = product of base and altitude divided by 2.

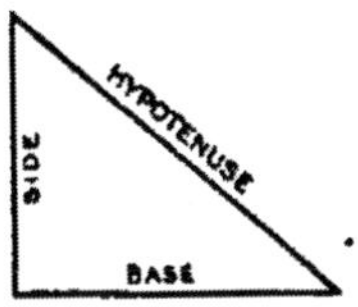

RIGHT TRIANGLE

A right triangle is one having one right angle.

The hypotenuse is the side opposite the right angle.

The square of the hypotenuse is equal to the sum of the squares of the other two sides.

$$\text{The area} = \frac{\text{base} \times \text{side}}{2}$$

$$\text{Hypotenuse} = \sqrt{\text{base squared} + \text{side squared.}}$$
$$\text{Base} = \sqrt{\text{hypotenuse squared} - \text{side squared.}}$$
$$\text{Side} = \sqrt{\text{hypotenuse squared} - \text{base squared.}}$$

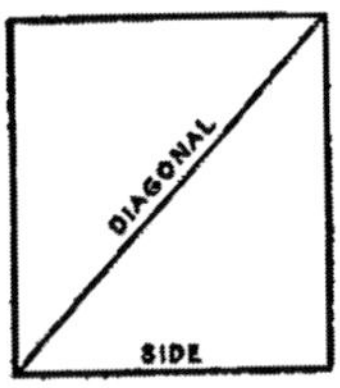

SQUARE

A plane figure of four sides. All four angles are right angles, and the opposite sides are equal and parallel. The sum of all the angles equals four right angles, or 360°.

$$\text{Area} = \text{square of a side.}$$
$$= \frac{\text{the square of a diagonal}}{2}$$

$$\text{Side} = \sqrt{\text{area}}$$
$$= \text{diagonal} \times .7071$$
$$\text{Diagonal} = \sqrt{\text{area}} \times 1.414$$
$$= \text{side} \times 1.414$$

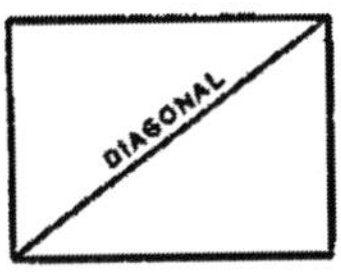

RECTANGLE

A plane figure of four sides. All four angles are right angles, and the opposite sides are equal and parallel. The sum of all the angles equals four right angles, or 360°.

The difference between a square and a rectangle

is that the adjacent sides of a square are equal; the adjacent sides of a rectangle need not be equal.

Area = product of two adjacent sides.
Short side = area divided by long side.
Long side = area divided by short side.
Diagonal = $\sqrt{\text{sum of squares of adjacent sides}}$.

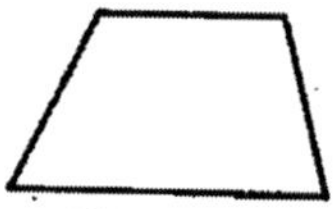

TRAPEZOID

A plane figure of four sides, two of which are parallel.

Area = sum of parallel sides × one-half the altitude.

REGULAR HEXAGON

A regular plane figure of six sides. All the sides are equal and all the angles are equal. The sum of all the angles equals 720°.

Area = square of side × 2.598.

Area = square of radius of circumscribed circle × 2.598.

Area = square of radius of inscribed circle × 3.464.

Side = radius of circumscribed circle.
Side = radius of inscribed circle × 1.155.
Radius of inscribed circle = side × .866.

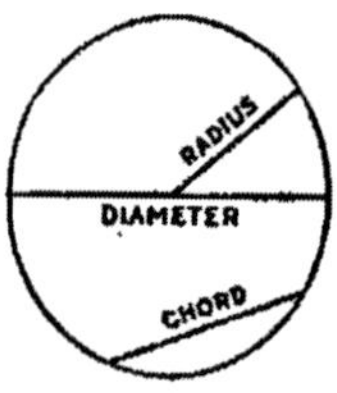

CIRCLE

A plane figure bounded by a curved line, every point of which is equally distant from a point within called the center.

A diameter is any straight line passing through the center and touching the circumference at each end.

Two circles having equal radii are equal.

Two circles with unequal radii vary in area as the squares of the radii—the circumferences are proportional to the radii.

A chord is a straight line intersecting or touching the circumference, but not passing through the center.

A chord at right angles to a diameter is divided into two equal parts by the diameter.

Circumference = diameter × 3.1416.
Area = square of radius × 3.1416.
Area = square of diameter × .7854.
Radius = circumference ÷ 6.2832.
Radius = $\sqrt{\text{area} \div 3.1416}$.

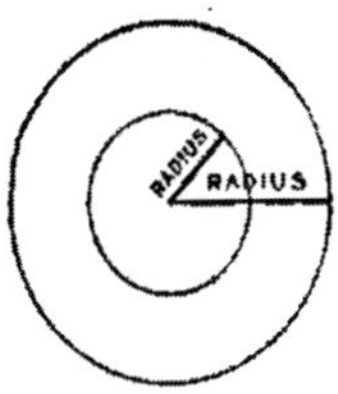

RING

A plane figure included between two circumferences having the same center.

Area = 3.1416 × (large radius squared − small radius squared).

Area = .7854 × (large diameter squared − small diameter squared).

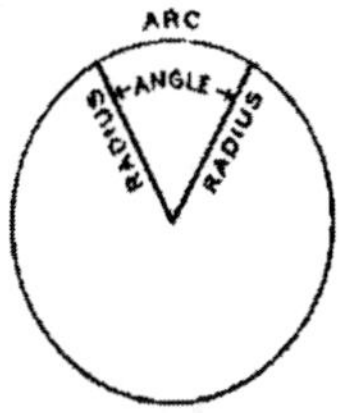

SECTOR

A plane figure included between two radii and the arc.

Area = one-half the radius × length of arc.

Area = .008727 × radius squared × angle in degrees.

$$\text{Angle} = \frac{57.296 \times \text{length of arc}}{\text{radius}}$$

$$\text{Radius} = \frac{57.296 \times \text{length of arc}}{\text{degrees in angle}}$$

Length of arc = .01745 × radius × degrees in angle.

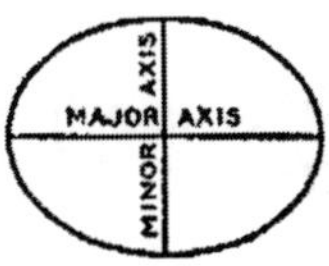

ELLIPSE

A plane figure bounded by a curve, of which every point is the same distance from two points on the longest axis; that is, the sum of the distances from any point to the foci is equal to the sum of the distances from any other point to the foci.

Area = 3.1416 × the product of its semi-axes.
Area = .7854 × product of axes.

$$\text{Circumference (approx.)} = 3.1416 \sqrt{\frac{\text{sum of square of axes}}{2}}$$

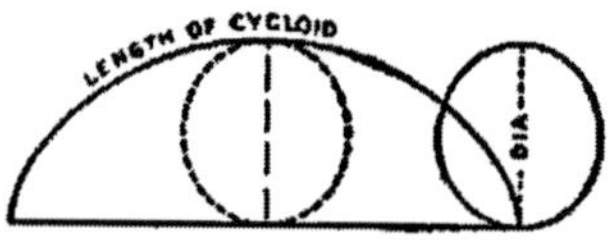

CYCLOID

A cycloid is a curve formed by a given point on a circumference of a circle rolling on a straight line.

Length of curve = diameter of circle × 4.
Length of curve = radius of circle × 8.
Area = 3 × 3.1416 × radius squared.
Area = 9.4248 × radius squared.
Area = area of circle × 3.

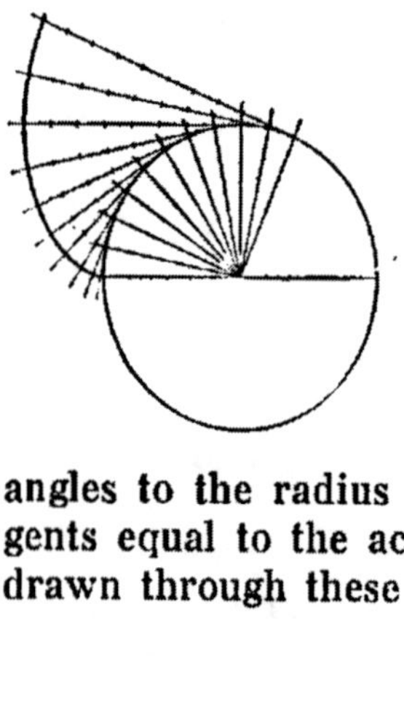

An involute is a curve traced by the end of a string as it unwinds from a cylinder and is kept taut. The string is always tangent to the cylinder. To draw the curve, divide the circumference into any number of equal parts, the smaller the number, the more accurate the curve. Through these points on the circumference, draw lines at right angles to the radius and make the lengths of these tangents equal to the actual length of the arcs. The curve drawn through these points is an involute.

SOLIDS

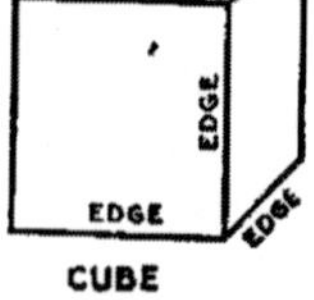

CUBE

A solid having six faces, each a square. All faces and edges are equal.
Volume = cube of edge.
Edge = $\sqrt[3]{\text{volume}}$.
Total area = square of edge × 6.

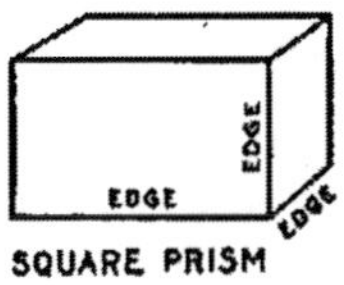

SQUARE PRISM

A solid having a rectangular base and rectangular sides. All opposite edges are equal and parallel.

Volume = product of the three edges.

Any edge = volume ÷ product of other two edges.

Total area = area of base and top + area of sides.

Total area = sum of areas of the six faces, all rectangular.

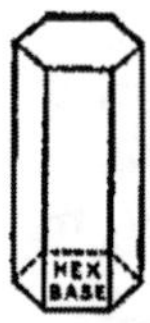

HEXAGONAL RIGHT PRISM

A prism having for its base a regular hexagon, and bases at right angles to faces.

Volume = 2.598 × square of side of base × vertical edge, or altitude.

Lateral area = side of base × vertical edge × 6.

Total area = lateral area + (5.196 × square of side of base).

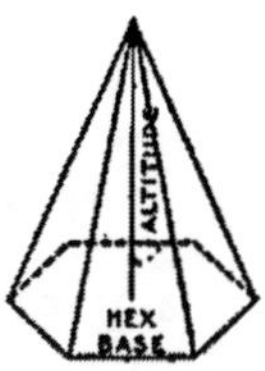

REGULAR PYRAMID

A right pyramid is a solid having a base a regular polygon and faces isosceles triangles.

Volume = one-third altitude × area of base.

Lateral area = perimeter of base × one-half slant height.

Slant height = altitude of triangular face.

Slant height = $\sqrt{\text{vertical edge squared} - \text{one-half side of base squared}}$.

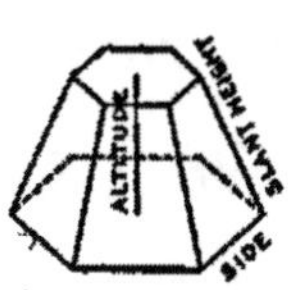

FRUSTUM OF PYRAMID

A frustum of a regular pyramid has parallel bases; that is, it is the lower portion of a pyramid cut by a plane parallel to the base.

Volume = sum of areas of the two bases and mean proportional between them × one-third altitude.

The mean proportional is equal to the square root of the product.

Lateral area = the sum of the perimeters of the two bases × one-half slant height.

Slant height = $\sqrt{\text{square of edge} - \text{square of one-half difference of side of bases}}$.

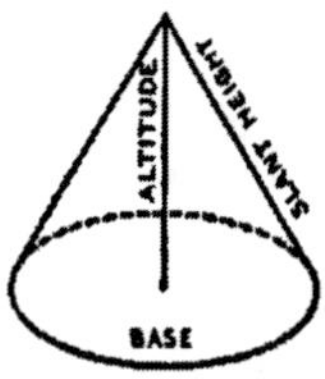

RIGHT CONE

A right cone has a circular base and vertex in a line perpendicular to the center of the base. It is a solid of revolution; that is, it is a solid figure formed by revolving a right triangle on its vertical side as an axis.

Volume = 1.0472 × square of radius of base × altitude.

Volume = .2618 × square of diameter of base × altitude.

Conical area = 3.1416 × radius of base × slant height.

Slant height = $\sqrt{\text{square of radius} + \text{square of altitude}}$.

Altitude = $\sqrt{\text{square of slant height} - \text{square of radius}}$.

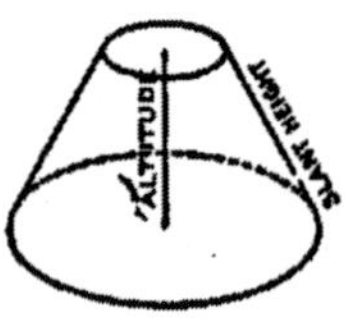

FRUSTUM OF CONE

The frustum of a cone has parallel bases. It is the lower portion of a cone when cut by a plane parallel to the base.

Volume = one-third altitude × sum of the areas of the two bases and the mean proportional between the two bases.

The mean proportional is equal to the square root of the product.

Lateral area = sum of perimeters (circles) of two bases × one-half slant height.

Slant height = $\sqrt{\text{square of altitude} + \text{square of difference in radii}}$.

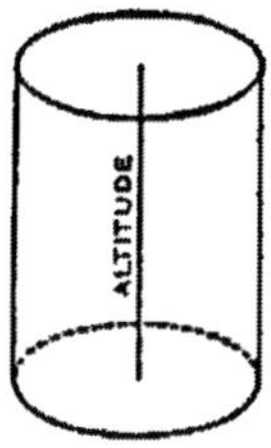

RIGHT CYLINDER

A right cylinder is a solid having circles for bases and lateral surface perpendicular to bases. It is a solid of revolution; that is, it is generated by revolving a rectangle about a side as an axis.

Volume = 3.1416 × square of radius × altitude.

Volume = .7854 × square of diameter × altitude.

Cylindrical surface = 6.2832 × radius × altitude.
Cylindrical surface = 3.1416 × diameter × altitude.

Total surface = cylindrical surface + twice area of (circle) base.

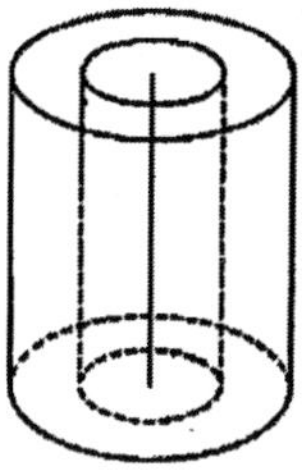

HOLLOW CYLINDER

Hollow cylinder; axis of hole coinciding with axis of cylinder.

Volume = difference in volume of two cylinders.

Volume = 3.1416 × altitude × (square of large radius − square of small radius).

Volume = 3.1416 × altitude × thickness × (large diameter − thickness).

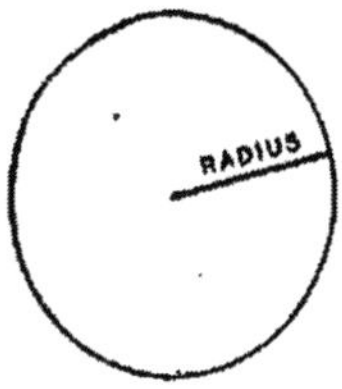

SPHERE

A sphere is a solid bounded by a curved surface every point of which is equally distant from a point within, called the center. It is a solid of revolution; that is, it is generated by revolving a half circle on the diameter as an axis.

$$\text{Volume} = \frac{4 \times 3.1416 \times \text{cube of radius}}{3}$$

$$= 4.1888 \times \text{cube of radius}$$

$$\text{Radius} = \sqrt[3]{\frac{\text{volume}}{4.1888}}$$

$$= .6204 \times \sqrt[3]{\text{volume.}}$$

Area = 4 × 3.1416 × square of radius,
= 12.5664 × square of radius.

$$\text{Radius} = \sqrt{\frac{\text{area}}{12.5664}}$$

$$= 3.5447 \times \sqrt{\text{area}}$$

Hollow sphere.

Volume = difference in volumes of two spheres.

Volume = 4.1888 × (cube of large radius − cube of small radius).

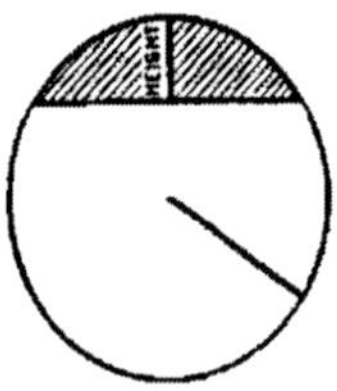

SPHERICAL SEGMENT

A spherical segment is formed by passing a plane through a sphere. If the plane passes through the center, the segment is one-half the sphere. If it does not pass through the center —

Volume = 3.1416 × square of height × (radius − one-third height).

Radius of segment = $\sqrt{\text{height} \times (\text{diameter of sphere} - \text{height of segment})}$.

Surface of spherical segment = 2 × 3.1416 × radius of sphere × height.

Surface of spherical segment = 6.2832 × radius of sphere × height.

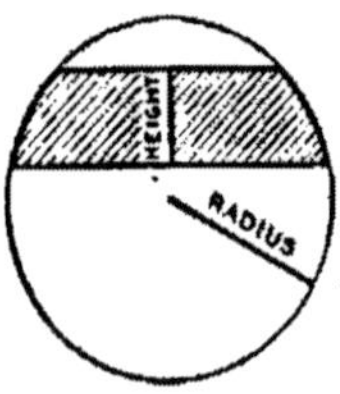

SPHERICAL ZONE

A spherical zone is formed by passing two parallel planes through a sphere.

Volume = volume of sphere − volume of segment.

Area = 2 × 3.1416 × radius of sphere × height.

Area = 6.2832 × radius of sphere × height.

MECHANICS

A FORCE is any cause which tends to produce or modify motion. It is measured in pounds, usually. Force has three characteristics — direction, place of application, magnitude.

WORK is the product of force and distance. It is measured in foot-pounds or in inch-pounds. Work does not involve the element time.

POWER is the amount of work done in a given time. It is the product of force and distance divided by time; and is expressed in foot-pounds per minute, or foot-pounds per second. The element of time is always included.

Power should not be given the same meaning as force, although some carelessly refer to an applied force as being a power.

VELOCITY is rate of motion. It is distance divided by time, and is expressed in feet per minute or feet per second. Velocity does not include force nor weight.

MOMENT OF FORCE. The moment of a force is the force multiplied by the perpendicular distance from the fixed point to the direction of the force. The fixed point is called the center of moments, and the perpendicular distance is called the lever arm of the force. Moment of force is measured in foot-pounds or inch-pounds.

GRAPHICAL REPRESENTATION OF FORCES. A force may be represented graphically by a straight line, the length being proportional to the magnitude. That is, the line is drawn to some scale. One end of the line represents the point of application, and an arrow head at the other end represents the direction.

Two or more forces may act together on a body.

To find a single force which produces the same effect as two or more forces, is to find the RESULTANT. The operation is called the COMPOSITION OF FORCES.

To find two or more forces which combined are equivalent to a given force is to find the COMPONENTS. The operation is called the RESOLUTION OF FORCES.

PARALLELOGRAM OF FORCES. When two forces acting at a point can be represented in direction and magnitude by the adjacent sides of a parallelogram, the resultant will be represented in direction and magnitude by the diagonal of the parallelogram. A B and A C are the forces and A R the resultant.

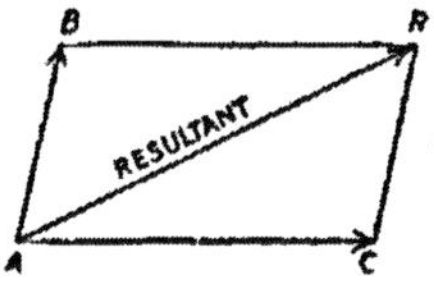

If two forces act in the same direction, the resultant is equal to their sum.

If two forces act in opposite directions, the resultant is their difference.

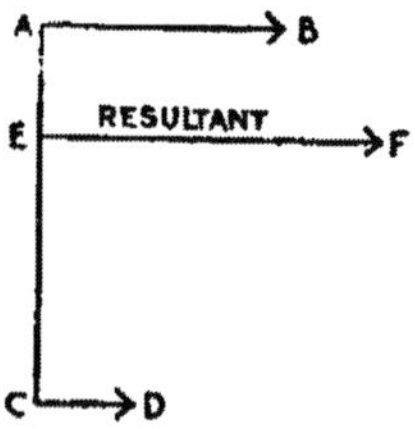

PARALLEL FORCES. When two forces are parallel and act in the same direction, but not from the same point, their resultant is parallel to both, and is equal to their sum. The resultant is located between the forces at a point that divides the line joining the points of application inversely as the magnitudes.

$$CD : AB = AE : EC$$

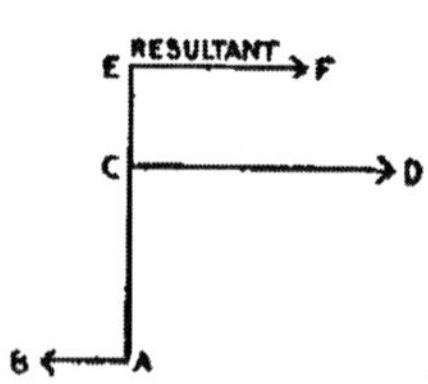

If the forces act in opposite directions, the resultant is parallel to both, but is located outside of them on the line (produced) joining the points of application. It is nearer the greater force and takes the same direction as the greater force, but in intensity it is equal to the difference between the components. The point of application of the resultant is:

$$AB : CD = CE : AE$$

LEVERS

Moments of forces are very important factors in machines. They may be illustrated in levers.

A lever is an inflexible rod, which may move about a fixed point called the fulcrum. The lever arms are the portions between the weights or forces and the fulcrum.

To solve all problems relating to the lever, it must be remembered that the moments are the weights or forces multiplied by the distances from the fulcrum; that is, by the lever arms.

As the lever is considered in balance, the product of the weight and length of weight arm is equal to the product of the power and length of power arm.

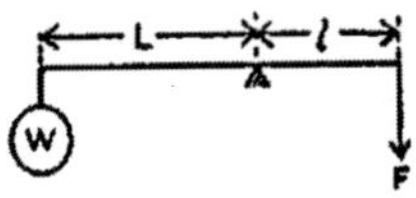

When the fulcrum is between the weight and the force, and both weight and force act in the same direction:

$$W \times L = F \times l$$

$$\text{or } W : F = l : L$$

$$W = \frac{F \times l}{L} \qquad L = \frac{F \times l}{W}$$

$$F = \frac{W \times L}{l} \qquad l = \frac{W \times L}{F}$$

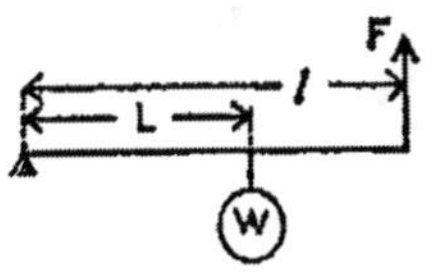

When the weight or load is between the fulcrum and the point at which the force is applied, the same principles apply; in fact, the same formulas are used.

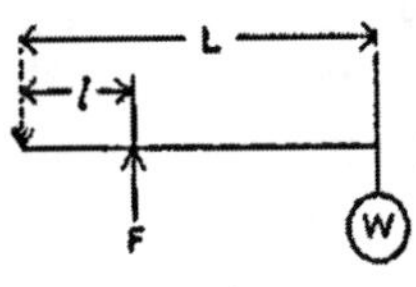

In the third form of lever, the force is applied at a point between the fulcrum and the weight. The same formulas are used.

If the weight of the lever itself is to be considered, the moment of force ($F \times l$) remains the same, but there are then several moments of weight. The additional moments of weight are found by multiplying the weight of the lever arm by the distance of its center of gravity from the fulcrum. In a lever of the first class there will be two moments of weight due to the weight of the lever, one will act with the moment of force and the other act with the moment of weight. With levers of the second and third class, the additional moment of weight will act with the original moment of weight, and, therefore, is added to it.

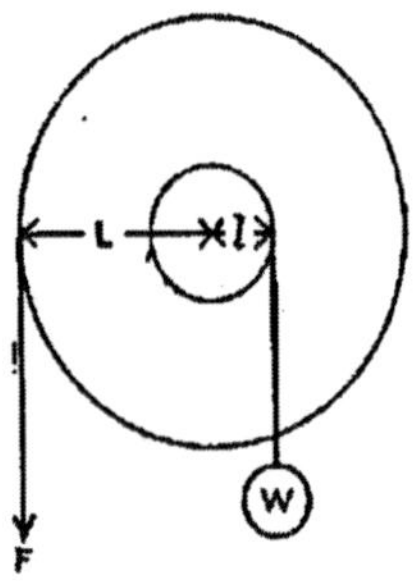

THE WINDLASS. The moment of force and the moment of weight are the means for finding the force required to lift a weight by a rope wound on the drum of a windlass.

$$F \times L = W \times l$$

$$F = \frac{W \times l}{L}$$

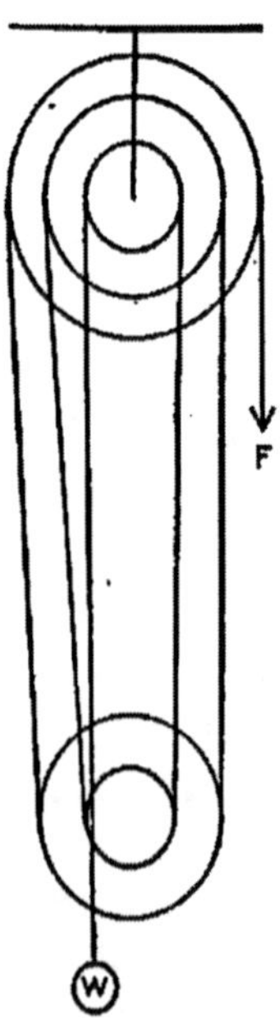

PULLEYS OR BLOCKS. The force required to lift the weight is equal to the weight divided by the number of ropes that are shortened.

$$F = \frac{W}{N}$$

If there are five ropes and the weight is 300 pounds, the force is:

$$F = \frac{300}{5} = 60 \text{ pounds}$$

The velocity with which the weight is raised is equal to the velocity of the force divided by the number of ropes shortened.

$$\text{Velocity} = \frac{\text{Velocity of F}}{N}$$

PULLEYS

A simple way to transmit power, either at the same speed, or a change of speed, is to place a pulley on the driving shaft and another on the driven shaft and pass an endless belt over them. It is evident that the linear speed of the pulleys is the same; that is, one revolution of the driving pulley pulls the belt through a distance equal to its circumference, and a point on the periphery of the driven pulley will be pulled through this distance whether or not the periphery is equal to the circumference of the driving pulley.

To change the rotative speed of shafts it is only necessary to place on them pulleys of unlike diameters.

The revolutions are inversely proportional to the circumferences and, therefore, to the diameters. The smaller pulley runs at the higher rotative speed.

D = diameter of driver.
d = diameter of driven.

Revs. of driven : Revs. of driver = D : d.
Revs. of driven × d = Revs. of driver × D.

The product of the revolutions and diameter of one pulley is equal to the product of the revolutions and diameter of the other pulley.

From Revs. of driven × d = Revs. of driver × D

$$\text{we have } d = \frac{\text{Revs. of driver} \times D}{\text{Revs. of driven}}$$

$$\text{and} \quad D = \frac{\text{Revs. of driven} \times d}{\text{Revs. of driver}}$$

To find the diameter of the driven pulley, multiply the revolutions of the driver by its diameter and divide by the revolutions of the driven.

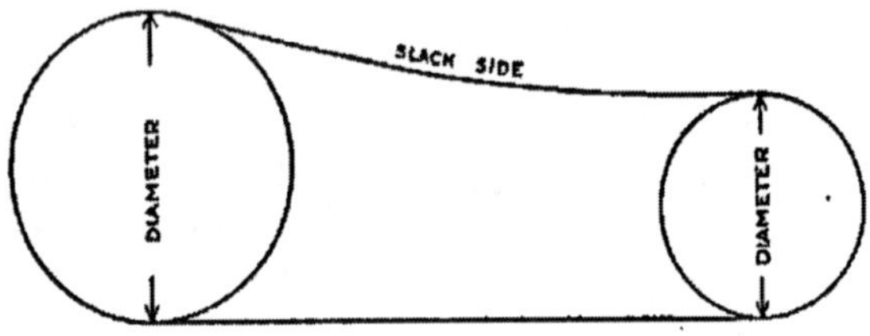

Example: The driving shaft makes 150 revolutions per minute and the driving pulley is 12 inches in diameter. The driven shaft is to make 600 revolutions; what diameter pulley should be selected?

$$d = \frac{150 \times 12}{600} = 3 \text{ inches}$$

The driving shaft makes 200 revolutions and the driven shaft is to make 150 revolutions per minute. With a driven pulley of 24 inches diameter, what size driver pulley should be used?

$$D = \frac{150 \times 24}{200} = 18 \text{ inches}$$

To find speeds when sizes of pulleys are known:

Revs. of driver $\times$ D = Revs. of driven $\times$ d.

$$\text{Revs. of driver} = \frac{\text{Revs. of driven} \times d}{D}$$

$$\text{Revs. of driven} = \frac{\text{Revs. of driver} \times D}{d}$$

Example: The driver pulley is 16 inches diameter and the driven is 18 inches diameter. When the driver runs at 270 revolutions per minute, what will be the speed of the driven pulley?

$$\text{Revs. of driven} = \frac{\text{Revs. of driver} \times D}{d}$$

$$= \frac{270 \times 16}{18} = 240$$

Example: Two pulleys, one of 14 inches diameter and the other of 18 inches diameter, are available. The driven shaft is to run at 120 revolutions per minute. If the 14-inch pulley is placed on the driven shaft what should be the speed of the driver?

$$\text{Revs. of driver} = \frac{\text{Revs. of driven} \times d}{D}$$

$$= \frac{120 \times 14}{18} = 93\ 1\text{-}3$$

FORMULAS FOR PULLEY DIAMETERS AND REVOLUTIONS

When three factors are known the fourth can be found by using one of the following formulas:

$$\text{Dia. of Driver} = \frac{\text{Dia of driven} \times \text{Revs. of driven}}{\text{Revs. of driver}}$$

$$\text{Dia. of Driven} = \frac{\text{Dia. of driver} \times \text{Revs. of driver}}{\text{Revs. of driven}}$$

$$\text{Revs. of Driver} = \frac{\text{Dia. of driven} \times \text{Revs. of driven}}{\text{Dia. of driver}}$$

$$\text{Revs. of Driven} = \frac{\text{Dia. of driver} \times \text{Revs. of driver}}{\text{Dia. of driven}}$$

The same principles apply to more complex belting. Suppose two pulleys are on the same shaft; we then have a combination that resembles a train of gears.

This arrangement is often desirable when it is impracticable to get the speed reduction with one belt; that is, when the larger pulley would have to be very large as compared with the smaller.

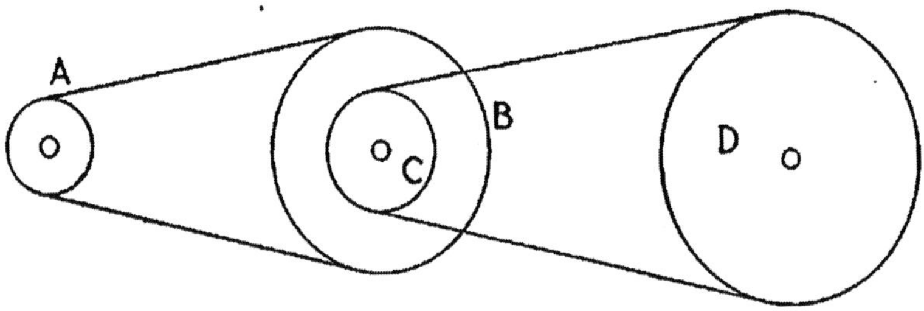

In the above illustration the high rotative speed of pulley A (on a motor shaft for example) is reduced to a much lower figure at pulley D.

Revs. of A × diameter of A = Revs. of B × diameter of B and Revs. of C × diameter of C = Revs. of D × diameter of D. But pulleys B and C are on the same shaft and have the same rotative speed.

Revs. of B = Revs. of C.

Combining these equations we may express the relation as follows:

The speed of the first driver multiplied by the diameters of all the drivers is equal to the speed of the last driven pulley multiplied by the diameters of all driven pulleys. Or

Revs. of A × diameter of A × diameter of C =
Revs. of D × diameter of B × diameter of D.

If five of the above quantities are known the sixth is easily found.

Example: Pulley A runs at 1200 Rev. per minute, and is 4 inches in diameter. Pulley B is 12 inches in diameter, C is 5 inches, and D is 16 inches. What is the speed of D?

$$1200 \times 4 \times 5 = \text{Revs. of D} \times 12 \times 16$$
$$24{,}000 = \text{Revs. of D} \times 192$$
$$\text{Revs. of D} = \frac{24{,}000}{192}$$
$$= 125$$

In the above we have found the rotative speed of D without finding the rotative speed of B, but we had given the diameters of B and C.

Suppose we had given the speed of D, but do not know what pulleys to use in place of B and C.

$$\frac{\text{Revs. of first driver}}{\text{Revs. of last driven}} = \frac{\text{product of diameters of all drivens}}{\text{product of diameters of all drivers}}$$

$$\text{or} \quad \frac{\text{Revs. of A}}{\text{Revs. of D}} = \frac{\text{diameter of B} \times \text{diameter of D}}{\text{diameter of A} \times \text{diameter of C}}$$

The two unknown quantities are diameter of B and diameter of C; but the RATIO can be found. Using the data in the above example we have

$$\frac{1200}{125} = \frac{16 \times \text{diameter of B}}{\text{diameter of C} \times 4}$$

$$\frac{\text{Diameter of B}}{\text{Diameter of C}} = \frac{4}{16} \times \frac{1200}{125}$$
$$= \frac{12}{5}$$

Then the ratio of the diameters is 12 : 5, and any pulleys having diameters in this ratio will give the desired speeds. The pulleys may be 12 and 5 inches, 18 and 7½, or 24 and 10.

Example: The shaft of 3-inch pulley D is to make 900 revolutions; what pulleys must be placed as B and C if A is 14 inches in diameter and makes 150 revolutions? The available pulleys have these diameters — 8, 9, 10½, 11, 12, 13½ inches.

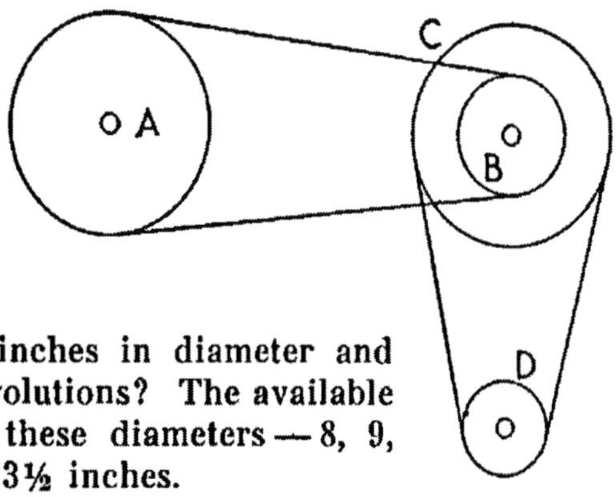

The formula to use is

$$\frac{\text{Revs. of first driver}}{\text{Revs. of last driven}} = \frac{\text{product of diameters of all drivens}}{\text{product of diameters of all drivers}}$$

$$\frac{150}{900} = \frac{\text{diameter of B} \times 3}{14 \times \text{diameter of C}}$$

$$\frac{1}{6} = \frac{3}{14} \times \frac{\text{diameter of B}}{\text{diameter of C}}$$

$$\frac{\text{Diameter of B}}{\text{Diameter of C}} = \frac{1}{6} \times \frac{14}{3}$$

$$= \frac{14}{18} = \frac{7}{9}$$

Then multiply the ratio 7 : 9 by any number which will make 7 and 9 equal to the diameters of pulleys on hand. Multiplying by 1½ gives 10½ and 13½.

To prove that the calculation is correct, place these values in this expression:

The speed of the first driver (150) multiplied by the diameters of all drivers (14) and (13½) is equal to the speed of the last driven (900) multiplied by the diameters of all driven pulleys (10½) and (3).

$$150 \times 14 \times 13\tfrac{1}{2} = 900 \times 10\tfrac{1}{2} \times 3$$
$$28{,}350 = 28{,}350$$

LENGTH OF BELTS

Open Belt. Pass a tape, preferably a steel tape, around the pulleys. This will give the length direct, if a single belt; but if a double belt is to be used add to the measurement twice the thickness of the belt. The length of small belts may be obtained by passing the belt around the pulleys and straining with hand pull.

New belts stretch and become slack after a short time, and the slack should be taken up. With long belts stretching may be anticipated by cutting the belt one inch shorter for every ten feet.

Rule for Length of Open Belt

Add diameters of pulleys in inches and multiply the sum by 1.57, then add to this product twice the distance between centers in inches.

Formula for Length of Open Belt

$$L = 3.14\,(R + r) + 2\,D + \frac{(R - r)^2}{D}$$

R = Radius of large pulley, inches.
r = radius of small pulley, inches.
D = Distance between centers of shaft, inches.
L = Length of belt, inches.

Formula for Length of Crossed Belt

$$L = 3.14\,(R + r) + 2\,D + \frac{(R + r)^2}{D}$$

The letters have the same values as above.

Example: Two pulleys are 11 feet apart and are 24 and 16 inches in diameter. Length of belt? Open and crossed.

$$L = 3.14 \times (12 + 8) + (2 \times 132) + \frac{(12 - 8)^2}{132}$$

$$= 62.8 + 264 + \frac{16}{132}$$

$$= 326.8 + .12$$

$$= 326.92 \text{ inches, open belt.}$$

$$L = 3.14 \times (12 + 8) + (2 \times 132) + \frac{(12 + 8)^2}{132}$$

$$= 62.8 + 264 + \frac{400}{132}$$

$$= 326.8 + 3$$

$$= 329.8 \text{ inches, crossed belt.}$$

GEARS

CONSTANT VELOCITY RATIO. Belts over pulleys and plain rolling cylinders cannot be depended upon to give a constant velocity ratio—there is always some loss of speed due to slip. But when two gears are in mesh a point on the pitch circle of one moves at the same linear velocity as a point on the pitch circle of the other, and the number of revolutions is always a constant ratio for these two gears.

Two gears in mesh have the same pitch; that is, the distance from the center of a tooth to the center of the next tooth, measured along the pitch circle, is the same for both gears. Therefore, two gears of the same pitch, but of different diameters, must have an unequal number of teeth.

It may be said that the space occupied by a tooth and the space between two teeth is the same in both gears if they have the same pitch. This fact shows immediately that the linear velocity of the pitch circles must be equal and the rotative speeds can be found in the same way as with belts. The pitch diameter or the number of teeth is substituted for the pulley diameter, for the numbers of teeth are proportional to the pitch diameters in the same way that the peripheries of pulleys are proportional to the diameters.

A gear having twice as many teeth as the gear meshing with it will make but one-half as many revolutions in a given time. Or, the speeds (rotative) are inversely as the number of teeth; the gear with the smaller number of teeth runs at the higher speed.

As in belts and pulleys, one gear of a pair is the driver and the other the driven or follower.

The number of revolutions of the driver multiplied by the number of teeth on the driver is equal to the number of revolutions of the follower multiplied by the number of teeth on the follower.

Revs. of driver × T = Revs. of follower × t, if T = number of teeth on the driver and t = number of teeth on the follower:

$$T = \frac{\text{Revs. of follower} \times t}{\text{Revs. of driver}}$$

$$\text{and} \quad t = \frac{\text{Revs. of driver} \times T}{\text{Revs. of follower}}$$

To find the number of teeth (T) on the driver, multiply the revolutions of the follower by its number of teeth and divide the product by the revolutions of the driver.

Example: The follower has 64 teeth and makes 30 revolutions per minute. The driver makes 80 revolutions per minute. How many teeth has the driver?

$$T = \frac{30 \times 64}{80} = 24$$

Example: The driver makes 160 revolutions per minute and has 40 teeth. The follower makes 100 revolutions. How many teeth?

$$t = \frac{160 \times 40}{100} = 64$$

$$\text{Revs. of driver} = \frac{\text{Revs. of follower} \times t}{T}$$

$$\text{Revs. of follower} = \frac{\text{Revs. of driver} \times T}{t}$$

Example: The follower has 90 teeth and makes 110 revolutions per minute. If the driver has 44 teeth, how many revolutions per minute?

$$\text{Revs. of driver} = \frac{110 \times 90}{44} = 225$$

Example: A driver having 63 teeth makes 800 revolutions per minute. If the follower has 42 teeth, what will be its speed?

$$\text{Revs. of follower} = \frac{800 \times 63}{42} = 1200$$

FORMULAS FOR SPEED OF GEARS

When three factors are known the fourth can be found by using one of the following formulas:

$$\text{Revs. of Driver} = \frac{\text{Revs. of follower} \times \text{teeth on follower}}{\text{teeth on driver}}$$

$$\text{Revs. of Follower} = \frac{\text{Revs. of driver} \times \text{teeth on driver}}{\text{teeth on follower}}$$

$$\text{Teeth on Driver} = \frac{\text{Revs. of follower} \times \text{teeth on follower}}{\text{Revs. of driver}}$$

$$\text{Teeth on Follower} = \frac{\text{Revs. of driver} \times \text{teeth on driver}}{\text{Revs. of follower}}$$

As in the case of pulleys, great speed changes are made by trains of gears in place of a pair. Examples are found in hoists, clocks, lathes, etc. Each pair in the train has its driver and follower, and if the shafts are parallel it is usual to get the speed change by keying two gears of unequal size on every shaft, except the first and last.

The velocity ratio of the first to the last is found as follows:

The product of the number of teeth on all the drivers divided by the product of the number of teeth on all the followers is the velocity ratio.

Suppose the train has three drivers, A, B, and C and three followers, L, M, and N.

A has 14 teeth and drives L having 70 teeth. Pinion B on same shaft with L has 13 teeth and drives M having 104 teeth. Pinion C has 15 teeth, and is on the same shaft with M; C drives N having 75 teeth. What is the velocity ratio of A to N?

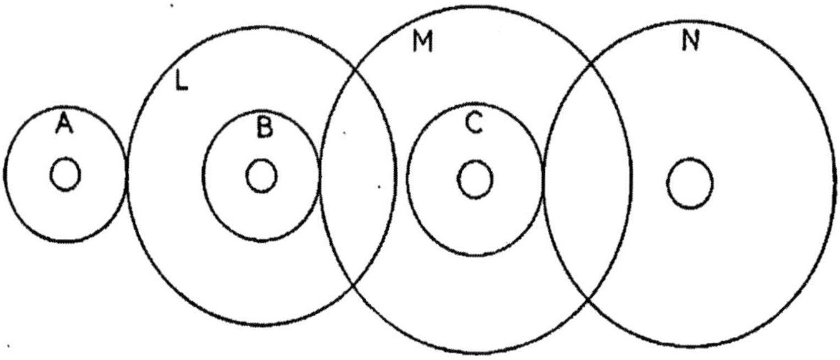

$$\text{Velocity ratio} = \frac{\text{teeth on A} \times \text{teeth on B} \times \text{teeth on C}}{\text{teeth on L} \times \text{teeth on M} \times \text{teeth on N}}$$

$$= \frac{14 \times 13 \times 15}{70 \times 104 \times 75}$$

$$= \frac{1}{200}$$

Knowing the velocity ratio of the train, it is easy to find the speed of N if the speed of A is known. If A runs at 1800 revolutions per minute, N will make only 9 revolutions for $1800 \div 200 = 9$.

When the speed of the first driver or the last follower is also known, the speed may be figured from the following:

Multiply the revolutions per minute of the first driver by the continued product of the number of teeth on all drivers, and divide by the continued product of the teeth on all followers. The quotient will be the revolutions per minute of the last follower.

LATHE GEARING

The apprentice who wishes to figure change gears for screw cutting should understand the *principles,* as already explained, rather than be dependent upon formulas. There is but one statement to be memorized.

The continued product of the speed of the first driver and the number of teeth on all drivers, is equal to the speed of the last follower multiplied by the continued product of the teeth on all followers.

In figuring change gears, the number of threads per inch to be cut corresponds to the revolutions of the driver, and the number of turns on the lead screw to move the carriage one inch corresponds to the speed of the follower.

Then the number of threads to be cut multiplied by the teeth on the spindle stud equals the number of threads on the lead screw multiplied by the teeth on the lead screw gear. Or

$$\frac{\text{threads to be cut}}{\text{threads on lead screw}} = \frac{\text{teeth on lead screw gear}}{\text{teeth on spindle stud}}$$

Suppose there are 6 threads on the lead screw and 46 teeth on the lead screw gear — how many threads will be cut if a 24-tooth gear is placed on the spindle stud?

$$\frac{\text{threads to be cut}}{6} = \frac{40}{24}$$

$$\text{threads to be cut} = \frac{40}{24} \times 6$$

$$= 10$$

The above assumes that the lathe is geared 1 : 1; that is, the lathe screw constant is equal to the number of threads per inch on the lead screw. If the lathe is not so geared, the lathe screw constant should be used in place of the threads per inch on the lead screw.

The foregoing example shows how the figuring can be done when the gears are on the spindle stud and lead screw; but the problem is usually one of finding out what gears to use.

Suppose seven threads are to be cut, and there are five threads per inch on the lead screw. What gears are to be used?

$$\frac{\text{threads to be cut}}{\text{threads on lead screw}} = \frac{\text{teeth on lead screw gear}}{\text{teeth on stud gear}}$$

$$\frac{7}{5} = \frac{\text{teeth on lead screw gear}}{\text{teeth on stud gear}}$$

The ratio of the gears is as 7 : 5.

By multiplying both 7 and 5 by any number, such as 6, we get

$$\frac{42}{30} = \frac{\text{teeth on lead screw gear}}{\text{teeth on stud gear}}$$

Using the formula as above may aid in disposing of that troublesome question, "Which gear goes on the stud?"

In some cases it may seem easier to assume one gear and go through the calculation to find the other, there being then one unknown quantity and three known quantities.

Table 13

Specific Gravity and Properties of Metals

Metal or Composition	Specific Gravity	Weight per Cubic Inch, Pounds	Melting Point. Deg. F.	Linear Expansion per Unit Length per Deg. F.
Aluminum	2.56	0.0924	1200	0.00001234
Antimony	6.71	0.2422	1150	0.00000627
Barium	3.75	0.1354	1560	
Bismuth	9.80	0.3538	500	0.00000975
Boron	2.60	0.0939		
Brass: 80 C., 20 Z.	8.60	0.3105	1700-1850	0.00000957
70 C., 30 Z.	8.40	0.3032		
60 C., 40 Z.	8.36	0.3018		
50 C., 50 Z.	8.20	0.2960		
Bronze	8.85	0.3195	1675	0.00000986
Cadmium	8.60	0.3105	610	
Calcium	1.57	0.0567	1450	
Chromium	6.50	0.2347	2740	
Cobalt	8.65	0.3123	2700	
Copper	8.82	0.3184	1940	0.00000887
Gold	19.32	0.6975	1930	0.00000786
Iridium	22.42	0.8094	4100	0.00000356
Iron, cast	7.20	0.2600	2300	0.00000556
Iron, wrought	7.85	0.2834	2900	0.00000648
Lead	11.37	0.4105	620	0.00001571
Magnesium	1.74	0.0628	1200	
Manganese	7.42	0.2679	2200	
Mercury (60° F.)	13.58	0.4902	—39	
Molybdenum	8.56	0.3090	4500	
Nickel	8.80	0.3177	2600	0.00000695
Platinum, rolled	22.67	0.8184	3200	0.00000479
Platinum, wire	21.04	0.7595		
Potassium	0.87	0.0314	144	
Silver	10.53	0.3802	1740	0.00001079
Sodium	0.98	0.0354	200	
Steel	7.80	0.2816	2500	0.00000636
Tellurium	6.25	0.2256	840	
Tin	7.29	0.2632	446	0.00001163
Titanium	3.54	0.1278	3360	
Tungsten	18.77	0.6776	5400	
Vanadium	5.50	0.1986	3200	
Zinc, cast	6.86	0.2476	785	0.00001407
Zinc, rolled	7.15	0.2581		

Table 14

Average Specific Gravity of Miscellaneous Substances

Substance	Specific Gravity	Weight per Cubic Foot, Lbs.
Asbestos	2.8	175
Asphaltum	1.4	87
Borax	1.75	109
Brick, common	1.8	112
Brick, fire	2.3	144
Brick, hard	2.0	125
Brick, pressed	2.15	134
Brickwork, in motor	1.6	100
Brickwork, in cement	1.8	112
Cement, Portland	3.1	194
Chalk	2.6	163
Charcoal	0.4	25
Coal, anthracite	1.5	94
Coal, bituminous	1.27	79
Concrete	2.2	137
Earth, loose	1.2	75
Earth, rammed	1.6	100
Emery	4.0	250
Glass	2.6	163
Granite	2.65	166
Gravel	1.75	109
Gypsum	2.2	137
Ice	0.9	56
Ivory	1.85	115
Limestone	2.6	163
Marble	2.7	169
Masonry	2.4	150
Mica	2.8	175
Mortar	1.5	94
Phosphorus	1.8	112
Plaster of Paris	1.8	112
Quartz	2.6	163
Salt, common	2.1	131
Sand, dry	1.6	100
Sand, wet	2.0	125
Sandstone	2.3	144
Slate	2.8	175
Soapstone	2.7	169
Soil, common black	2.0	125
Sulphur	2.0	125
Trap	3.0	187
Tile	1.8	112

Table 15

Specific Gravity of Gases

(At 32 degrees F.)

Gas	Sp. Gr.	Gas	Sp. Gr.
Air	1.000	Hydrogen	0.069
Acetylene	0.910	Illuminating gas	0.040
Alcohol vapor	1.601	Mercury vapor	6.940
Ammonia	0.592	Marsh gas	0.555
Carbon dioxide	1.520	Nitrogen	0.971
Carbon monoxide	0.967	Nitric oxide	1.039
Chlorine	2.423	Nitrous oxide	1.527
Ether vapor	2.586	Oxygen	1.106
Ethylene	0.967	Sulphur dioxide	2.250
Hydrofluoric acid	2.370	Water vapor	0.623
Hydrochloric acid	1.261		

1 cubic foot of air at 32 degrees F. and atmospheric pressure weighs 0.0807 pound

Table 16

Specific Gravity of Liquids

Liquid	Sp. Gr.	Liquid	Sp. Gr.
Acetic acid	1.06	Muriatic acid	1.20
Alcohol, commercial	0.83	Naphtha	0.76
Alcohol, pure	0.79	Nitric acid	1.22
Ammonia	0.89	Olive oil	0.92
Benzine	0.69	Palm oil	0.97
Bromine	2.97	Petroleum oil	0.82
Carbolic acid	0.96	Phosphoric acid	1.56
Carbon disulphide	1.26	Rape oil	0.92
Cotton-seed oil	0.93	Sulphuric acid	1.84
Ether, sulphuric	0.72	Tar	1.00
Fluoric acid	1.50	Turpentine oil	0.87
Gasoline	0.90	Vinegar	1.08
Kerosene	0.80	Water	1.00
Linseed oil	0.94	Water, sea	1.03
Mineral oil	0.92	Whale oil	0.92

Table 17
Composition of Miscellaneous Alloys

Alloys	Antimony	Bismuth	Copper	Iron	Lead	Nickel	Tin	Zinc
Brass, common yellow			61.6		2.9		0.2	35.3
Brass, to be rolled			32				1.5	10
Brass castings, common			20				2.5	1.25
Gun metal			8				1	
Copper flanges			9				0.26	1
Bronze Statuary			91.4		1.37		1.7	5.53
German Silver			2	6.5		7.9		6.3
Britannia metal	50	25					25	
Chinese white copper			20.2			15.8	1.3	12.7
Pattern letters	15	15			70			
Bell metal			4				1	
Chinese gongs			40.5				9.2	
White metal, ordinary	28.4		3.7				14.2	3.7
Spelter			1					1
Type metal	1				3–7			

Table 18

Average Specific Heats of Various Substances

Substance	Specific Heat	Substance	Specific Heat
Alcohol (absolute)	0.700	Kerosene	0.500
Alcohol (density 0.8)	0.622	Lead	0.031
Aluminum	0.214	Limestone	0.217
Antimony	0.051	Magnesia	0.222
Benzine	0.450	Marble	0.210
Brass	0.094	Masonry, brick	0.200
Brickwork	0.200	Mercury	0.033
Cadmium	0.057	Naphtha	0.310
Charcoal	0.200	Nickel	0.109
Chalk	0.215	Oil, machine	0.400
Coal	0.240	Oil, olive	0.350
Coke	0.203	Phosphorus	0.189
Copper, 32° to 212° F.	0.094	Platinum	0.032
Copper, 32° to 572° F.	0.101	Quartz	0.188
Corundum	0.198	Sand	0.195
Ether	0.503	Silica	0.191
Fusel oil	0.564	Silver	0.056
Glass	0.194	Soda	0.231
Gold	0.031	Steel, mild	0.116
Graphite	0.201	Steel, high carbon	0.117
Ice	0.504	Stone (generally)	0.200
Iron, cast	0.130	Sulphur	0.178
Iron, wrought, 32° to 212° F.	0.110	Sulphuric acid	0.330
32° to 392° F.	0.115	Tin	0.056
32° to 572° F.	0.122	Turpentine	0.472
32° to 662° F.	0.126	Water	1.000
Iron, at high temperatures:		Wood, fir	0.650
1382° to 1832° F.	0.213	Wood, oak	0.570
1750° to 1840° F.	0.218	Wood, pine	0.467
1920° to 2190° F.	0.199	Zinc	0.095

Table 19

Templets for Drilling Standard and Low Pressure Flanged Valves and Fittings—American Standard

Size	Diam. of Flange	Thickness of Flange	Diam. of Bolt Circle	No. of Bolts	Size of Bolts
1	4	7/16	3	4	7/16
1¼	4½	½	3⅜	4	7/16
1½	5	9/16	3⅞	4	½
2	6	⅝	4¾	4	⅝
2½	7	11/16	5½	4	⅝
3	7½	¾	6	4	⅝
3½	8½	13/16	7	4	⅝
4	9	15/16	7½	8	⅝
4½	9¼	15/16	7¾	8	¾
5	10	15/16	8½	8	¾
6	11	1	9½	8	¾
7	12½	1 1/16	10¾	8	¾
8	13½	1⅛	11¾	8	¾
9	15	1⅛	13¼	12	¾
10	16	1 3/16	14¼	12	⅞
12	19	1¼	17	12	⅞
14	21	1⅜	18¾	12	1
15	22¼	1⅜	20	16	1
16	23½	1 7/16	21¼	16	1
18	25	1 9/16	22¾	16	1⅛
20	27½	1 11/16	25	20	1⅛
22	29½	1 13/16	27¼	20	1¼
24	32	1⅞	29½	20	1¼
26	34¼	2	31¾	24	1¼
28	36½	2 1/16	34	28	1¼
30	38¾	2⅛	36	28	1⅜
32	41¾	2¼	38½	28	1½
34	43¾	2 5/16	40½	32	1½
36	46	2⅜	42¾	32	1½
38	48¾	2⅜	45¼	32	1⅝
40	50¾	2½	47¼	36	1⅝
42	53	2⅝	49½	36	1⅝
44	55¼	2⅝	51¾	40	1⅝
46	57¼	2 11/16	53¾	40	1⅝
48	59½	2¾	56	44	1⅝
50	61¾	2¾	58¼	44	1¾
52	64	2⅞	60½	44	1¾
54	66¼	3	62¾	44	1¾
56	68¾	3	65	48	1¾
58	71	3⅛	67¼	48	1¾
60	73	3⅛	69¼	52	1¾
62	75¾	3¼	71¾	52	1⅞
64	78	3¼	74	52	1⅞
66	80	3⅜	76	52	1⅞
68	82¼	3⅜	78¼	56	1⅞
70	84½	3½	80½	56	1⅞
72	86½	3½	82½	60	1⅞
74	88½	3⅝	84½	60	1⅞
76	90¾	3⅝	86½	60	1⅞
78	93	3¾	88¾	60	2
80	95¼	3¾	91	60	2
82	97½	3⅞	93¼	60	2
84	99¾	3⅞	95½	64	2
86	102	4	97¾	64	2
88	104¼	4	100	68	2
90	106½	4⅛	102¼	68	2⅛
92	108¾	4⅛	104½	68	2⅛
94	111	4¼	106¼	68	2⅛
96	113¼	4¼	108½	68	2¼
98	115½	4⅜	110¾	68	2¼
100	117¾	4⅜	113	68	2¼
....	....	....	....	....	....

Bolt holes are drilled ⅛ inch larger than nominal diameter of bolts.

Table 20

Templets for Drilling Extra Heavy Flanged Valves and Fittings—American Standard

Size	Diam. of Flange	Thickness of Flange	Diam. of Bolt Circle	No. of Bolts	Size of Bolts
1	4 1/2	11/16	3 1/4	4	1/2
1 1/4	5	3/4	3 3/4	4	1/2
1 1/2	6	13/16	4 1/2	4	5/8
2	6 1/2	7/8	5	4	5/8
2 1/2	7 1/2	1	5 7/8	4	3/4
3	8 1/4	1 1/8	6 5/8	8	3/4
3 1/2	9	1 3/16	7 1/4	8	3/4
4	10	1 1/4	7 7/8	8	3/4
4 1/2	10 1/2	1 5/16	8 1/2	8	3/4
5	11	1 3/8	9 1/4	8	3/4
6	12 1/2	1 7/16	10 5/8	12	3/4
7	14	1 1/2	11 7/8	12	7/8
8	15	1 5/8	13	12	7/8
9	16 1/4	1 3/4	14	12	1
10	17 1/2	1 7/8	15 1/4	16	1
12	20 1/2	2	17 3/4	16	1 1/8
14	23	2 1/8	20 1/4	20	1 1/8
15	24 1/2	2 3/16	21 1/2	20	1 1/4
16	25 1/2	2 1/4	22 1/2	20	1 1/4
18	28	2 3/8	24 3/4	24	1 1/4
20	30 1/2	2 1/2	27	24	1 3/8
22	33	2 5/8	29 1/4	24	1 1/2
24	36	2 3/4	32	24	1 5/8
26	38 1/4	2 13/16	34 1/2	28	1 5/8
28	40 3/4	2 15/16	37	28	1 5/8
30	43	3	39 1/4	28	1 3/4
32	45 1/4	3 1/8	41 1/2	28	1 7/8
34	47 1/2	3 1/4	43 1/2	28	1 7/8
36	50	3 3/8	46	32	1 7/8
38	52 1/4	3 7/16	48	32	1 7/8
40	54 1/2	3 9/16	50 1/4	36	1 7/8
42	57	3 11/16	52 3/4	36	1 7/8
44	59 1/4	3 3/4	55	36	2
46	61 1/2	3 7/8	57 1/4	40	2
48	65	4	60 3/4	40	2

Bolt holes are drilled 1/8 inch larger than nominal diameter of bolts.

Table 21—Tap Drills
For A. S. M. E. Standard and Special Machine Screw Taps

The diameter given for each hole to be tapped allows for a practical clearance at the root of the thread of the screw and will not impose undue strain upon the tap in service.

Size of Tap	No. of Threads	Size of Drill	Size of Tap	No. of Threads	Size of Drill
0	80	.0465	9	32	.1405
1	**64**	**.055**	**10**	**24**	**.140**
1	72	.0595	10	30	.152
2	**56**	**.0670**	**10**	**32**	**.154**
2	64	.070	**12**	**24**	**.166**
3	**48**	**.076**	12	28	.173
3	56	.0785	**14**	**20**	**.182**
4	**36**	**.080**	14	24	.1935
4	**40**	**.082**	**16**	**20**	**.209**
4	48	.089	16	22	.213
5	**36**	**.0935**	**18**	**18**	**.228**
5	**40**	**.098**	18	20	.234
5	44	.0995	**20**	**18**	**.257**
6	**32**	**.1015**	20	20	.261
6	**36**	**.1065**	**22**	**16**	**.272**
6	40	.110	22	18	.281
7	**30**	**.113**	24	16	.295
7	**32**	**.116**	**24**	**18**	**.302**
7	36	.120	**26**	**14**	**.316**
8	**30**	**.1285**	26	16	.323
8	**32**	**.1285**	28	14	.339
8	36	.136	**28**	**16**	**.348**
9	**24**	**.1285**	30	14	.368
9	**30**	**.136**	**30**	**16**	**.377**

NOTE:—Special Taps are in Bold Face Type.

Table 22—Tap Drills for Machine Screws

Size of Tap	American Standard Diameter in Inches	Size of Drill for Outside Diameter of Screw	Size of Drill for Tapping Hole
2 x 48 2 x 56 2 x 64	.25763	44	50 49 48
3 x 40 3 x 48 3 x 56	.22942	39	49 47 45
4 x 32 4 x 36 4 x 40	.20431	33	46 44 43
5 x 30 5 x 32 5 x 36 5 x 40	.18194	1/8	43 42 41 38
6 x 30 6 x 32 6 x 36 6 x 40	.16202	28	38 37 36 35
7 x 28 7 x 30 7 x 32	.14428	24	34 33 32
8 x 24 8 x 30 8 x 32	.12849	19	31 31 30
9 x 24 9 x 28 9 x 30 9 x 32	.11443	16	30 28 28 26
10 x 24 10 x 30 10 x 32	.10189	11	26 24 24
11 x 24 11 x 28 11 x 30	.090742	6	21 20 19
12 x 20 12 x 22 12 x 24 12 x 28	.080808	5/32	24 20 19 18

Size of Tap	American Standard Diameter in Inches	Size of Drill for Outside Diameter of Screw	Size of Drill for Tapping Hole
13 x 20 13 x 22 13 x 24	.071961	15/64	17 17 15
14 x 20 14 x 22 14 x 24	.064084	1/4	15 11 10
15 x 18 15 x 20 15 x 22 15 x 24	.057068	F	12 10 8 7
16 x 16 16 x 18 16 x 20	.05082	I	12 8 7
17 x 16 17 x 18 17 x 20	.045257	L	8 4 3
18 x 16 18 x 18 18 x 20	.040303	19/64	2 2 1
19 x 16 19 x 18 19 x 20	.03589	5/16	1 B C
20 x 16 20 x 18 20 x 20	.031961	P	C E F
22 x 16 22 x 18	.025347	S	H J
24 x 14 24 x 16 24 x 18	.0201	3/8	L M N
26 x 14 26 x 16	.01594	13/32	O P
28 x 14 28 x 16	.012641	7/16	R S
30 x 14 30 x 16	.010025	29/64	U V

INDEX

SETS OF TOOLS
FOR APPRENTICES AND STUDENTS
SET NO. 900
IN FOLDING LEATHER CASE

Size of case folded, 7″ x 4¾″ x 1⅜″

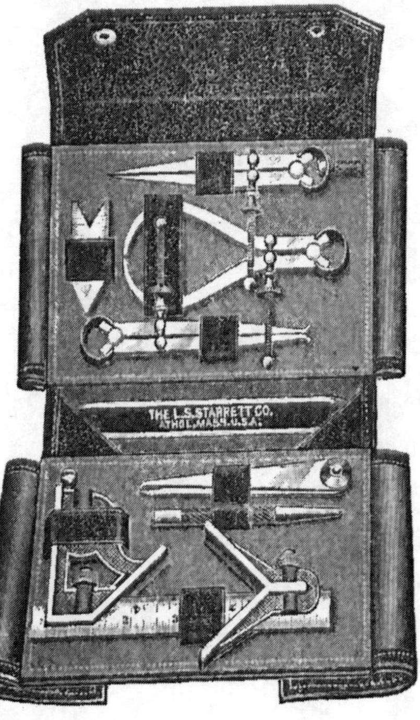

Set No. 900 consists of the leather case and the following tools:

No. 11, 6″ Combination Square, complete
No. 117B, Center Punch
No. 321, 6″ Flexible Steel Rule in pocket case
No. 390, Center Gage
No. 241, 4″ Caliper
No. 79, 4″ Outside Caliper with solid nut
No. 73, 4″ Inside Caliper with solid nut
No. 83, 4″ Divider with solid nut

PRICE, set complete $6.00

SETS OF TOOLS

FOR APPRENTICES AND STUDENTS

SET NO. 901

IN NICELY FINISHED WOODEN CASE

Size of case, 12″ x 7″ x 1½″

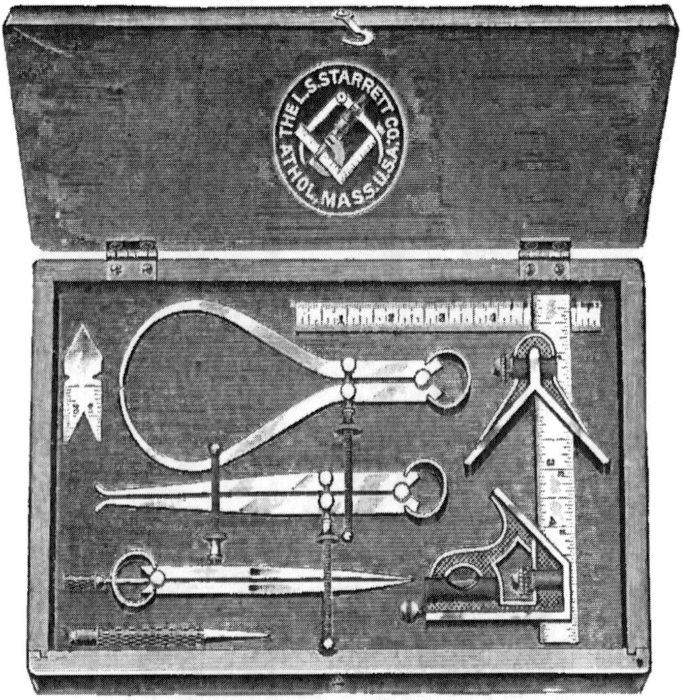

Set No. 901 consists of the wooden case and the following tools:

No. 11, 6″ Combination Square, complete
No. 321, 6″ Flexible Steel Rule in pocket case
No. 117B, Center Punch
No. 390, Center Gage
No. 77, 5″ Divider with solid nut
No. 79, 6″ Outside Caliper with solid nut
No. 73, 6″ Inside Caliper with solid nut

PRICE, set complete $6.15